TIME AND CONSCIOUSNESS

TIME AND CONSCIOUSNESS

Cyclical, Hierarchical, and Causal Notions of Time

Ashish Dalela

Time and Consciousness—Cyclical, Hierarchical, and
Causal Notions of Time
by Ashish Dalela

Published by Shabda Press
www.shabdapress.net
ISBN 978-93-85384-28-8

SHABDA
PRESS

Contents

List of Figures

Preface

Ticking away the moments that make up a dull day
You fritter and waste the hours in an off-hand way
Kicking around on a piece of ground in your hometown
Waiting for someone or something to show you the way
— *Roger Waters*

Questions about the nature of time have always been an important part of physics and philosophy, but they have never been answered satisfactorily. The simplest of these questions is the idea that time passes, but even though it passes, we have access to the past through memory, and to the future by imagination. All scientific ideas, which later become well-known, were at one time unknown. They were then perceived by the mind, and expressed through a scientific language, which is when others were able to understand them. In short, all that we know scientifically today also existed in the past, was perceived by someone at some time, and then converted into a reality that others can also see. Likewise, all of us carry memories of the past, which shape us in numerous ways. The past as memory, the future as goals, and the present as exigent circumstances, therefore, all shape our experience, and alter the present, the memory of the past, and the future possible goals. How this happens, involves a thus far scientifically unknown process.

Modern science, however, advances a simplistic idea: only the present exists; the future and the past are *physically* unreal. Whatever goals you imagine, or the pasts you remember, are but the epiphenomena of the chemicals in your brain. Since the chemicals do not exist in the past or the future, therefore, no new science based on the commonsensical ideas of the existence of the past or the future is required, because physical science based upon the idea that only the present exists, is causally sufficient. Of course, because there is no scientific explanation of the mind at the present, therefore, the claims of this materialistic view cannot be rationally justified, but they persist due to the deep desire to eliminate the mind from all science. If only the mind would disappear by the mind wishing itself away!

The future as goals, and the past as memories exist in our mind as concepts, and they cannot be reduced to physical properties unless someone can explain the emergence of abstraction, representation, intention, and reference—the fundamental properties by which we identify the mind. Materialism makes science a burlesque of rationality and experience by ignoring the immediate and pervasive aspects of our experience, and if that fails, then advocating the most convoluted excuses to avoid the obvious. In Indian philosophy, a person who finds the most obvious to be unreal and the most unreal to be obvious is said to be affected by the mode of 'darkness'.

On one hand, this caricature of reality blatantly contradicts experience, and what we consider to be true. On the other hand, if the past and future indeed influence the present, then all causal explanations based on the present alone must be incomplete. Many forms of incompleteness are already known in physical theories, but they are not connected to the absence of a causal role for the past and the future, due to the simplistic ideas about reality as that which exists at the present. And unless this connection is made in science, the theories will remain incomplete. Thus, the inability to understand the role of the past and the future presents not just human problems (i.e. that the physicalist doctrine contradicts our lived experience), but also scientific problems (i.e. that the theories of science are also incomplete).

The fact is that, for a conscious person, both the past and the future are causally efficacious. But in what form do they exist at the present? The short answer is that we must recognize their existence as a *possibility*. This possibility is accessible to some people, not to everyone, but that is no different than the fact that the things inside my bedroom are accessible to me, and not to everyone on this planet. To accommodate the existence of the past and the future at the present, we need to change our understanding of matter, namely, that it exists as a possibility. But this alone will not suffice for all causal scenarios because some such possibilities are accessible through our senses, while other possibilities are only accessible through the mind. Some of these possibilities are seen during waking, while others are seen during dreaming. Sometimes we can see things by the mind upon hearing its description in language, and sometimes we must see these things through our senses as physical things or models of reality before we can see them through the mind. Therefore, the simple idea that the world exists as a possibility, must be updated to recognize several *kinds* of possibilities.

What is accessible through the senses is also accessible through the mind, but what is accessible through the mind may not always be accessible through the senses. For example, you can feel another person's pain, but you cannot see the pain through the senses. Therefore, there is also a *hierarchy* among these types of possibilities, with the higher-level possibilities being accessible through the mind and the senses, while the lower-level possibilities are perceivable only through the senses. When multiple tiers of possibilities are recognized, then the evolution of these tiers of possibilities also creates many notions of time—applied to many tiers of possibilities.

There is hence the evolution of the body—from childhood, to youth, to old age, and death. There is an evolution of our sensual capacities—e.g. the development of greater refinement in musical abilities or craftsmanship. Then there is a mental evolution, such as the mastering of a language or a technical subject over time. Similarly, there is an intellectual evolution in our belief system, or what we consider to be true or false. Our goals, aspirations, and intentions also evolve with time. And sometimes, what we value—e.g. beauty vs. wealth vs. knowledge vs. power—also changes with time.

When we recall the past from memory, our notion of past can involve the idea that at one time I was small, that at one time I was learning to talk and walk, that at one time I did not understand algebra, that at that time I did not believe that I had to be responsible for my choices, that I had the ambitions to become a police officer or a fireman, or that I did not value hard work over play. Likewise, when we think of the future, we can have different ideas about what we want to be, in terms of our body, abilities, thoughts, beliefs, goals, and values. Thus, in our experience, time has many varieties, and since a certain type of body can co-exist with many types of skills, thoughts, beliefs, goals, and values, therefore, each of these can be said to be evolving differently and individually over time. With the hierarchy of realities in our experience—i.e. some realities being 'deeper' than the others—the time of our experience also becomes hierarchical. Similarly, because the deeper reality changes slowly, therefore, time is experienced to elapse at different rates at different levels. Thus, a person who doesn't understand that life must be led responsibly, is said to be mentally a 'child' although his body has grown into an adult. Someone who has naïve assumptions about the world is said to be 'childish' even in an adult body. If someone has simplistic goals, he is said to have a 'child-like innocence'.

Modern philosophical and scientific discourse on time is so far removed from the practical lived experience of time, that it has almost nothing concrete to tell us about it. Nowhere in science are so many obviously erroneous ideas about anything so prevalent and fashionable than in the understanding of time. Never has been a subject so intuitively accessible been presented in such non-intuitive and patently contrary to experience ways.

But there are some interesting ideas about time being a 'dimension', about time being an 'arrow', that time evolution is 'continuous', etc. that are worthy contrasts to the idea of time experience. It is safe to say that modern science understands so little about time that it equates time with space for all practical purposes. Just as particles move in space, similarly, the 'present' moves like a particle on the time axis; this motion is continuous like the movement of particles and space. The main perplexing problem of time in modern science is that time moves in one direction—i.e. from past to future—whereas particles on a space-like axis can move in both directions. Therefore, if you talk to a scientist, she might say that science is only aware of one problem regarding time, which is that science doesn't understand why time only moves in one direction, rather than both. The dissimilarities between space and time—for example, how a 'particle' moves along the time direction—continue to be ignored by the majority of scientists.

It is also typical of modern science to convert anything that we don't understand into a 'law' of nature, because at that point you can stop asking why this thing behaves in this way, and simply accept it as a *fact* of reality. In the case of time, the forward direction of time is treated as the "second law of thermodynamics", although questions about the reversibility of time in classical mechanics, and its irreversibility in thermodynamics remain. Classical mechanics operates on the assumption that nature is deterministic and only one out of all the possible states is real. Thermodynamics on the other hand operates on the principle that if we don't fix the initial conditions of a system, then a system alternates between all possible states it potentially can. Through a succession of kooks, the discussion now whittles down to how the evolution of heat is different than the motion of classical objects, and any semblance to the problems of the passing of time is lost. It is now heretical to go back to the original questions because most scientists will facetiously declare: "we are not interested in philosophical issues", disregarding the fact that the current scientific ideas of time are so detached from pervasive experience, that no progress in the understanding of time can be expected so long as the genie is kept bottled inside simplistic theories.

When time is presented as an 'axis', just like space, then we arrive at another counterintuitive idea that time passes *linearly* rather than *cyclically*. This view of time has more to do with Western culture than with anything seen in the real world. In the real world, for example, air, soil, and water create food grains, which we cook into palatable dishes, and the waste is recycled back into air, soil, and water. In the real world, the sun rises, sets, and rises again; the seasons change cyclically; the money goes round and round in an economic system; civilizations rise, fall, and rise again; the same political parties come into power and lose power again and again; and the same ideas come into and go out of fashion over and over. If there is anything pervasive about change, it is that history repeats itself over and over. The only thing that *seems* to go linearly is the progression from childhood to youth to old age and death. The cultures in the East believe that after every death is a rebirth. And the cultures in the West believe that death is the end of everything[1]. Based on these two cultural belief systems, time is viewed cyclically in the East, and whatever seems progressive is considered a small part of a slow and big cycle. The linearity is accepted, but it is not regarded as a fundamental idea about time. Inversely, in the West, whatever seems circular and cyclical is considered unnatural compared to what is linear.

This cultural bias about the world is ensconced in Newton's laws of motion. The first law says that objects move in a straight line if undisturbed by a force. In other words, the linear motion is the default state of reality. The second law says that objects change their trajectory due to a force. In other words, if something moves cyclically, then it must be due to a force. The thinking in the East is completely different. In Eastern thought, life follows death which follows life, and because everyone is stuck in the cycle of birth and death, therefore, this cyclical change is the normal thing. The exceptional thing is if we try to get out of this cycle and that escape from the cycle of birth and death constitutes the basic idea of progression.

This essential view of life pervades all philosophical and scientific thinking in the East. Since the default state of nature is cyclical, therefore, most progressions are the 'illusions' of a larger cycle. Hence, the changes of this world—such as the advancement of the materialistic civilization—is considered an illusion: It just *seems* progressive, but it is only a small part of a larger circle of life, and whatever is achieved today will be destroyed tomorrow. The real progress is when someone gets out of these cycles. Just as the West believes that to move things out of a straight-line path one must apply some force, similarly, the East thinks that to take something out of the periodic cyclical behaviors, some force must be applied. Thus, the West considers the creation of machines that move like flywheels and clocks an achievement, while the East views these as natural hamster wheels.

These differences in cultural outlook pervade the Western and Eastern theories of the world, their changes, and ultimately time. For example, in the West, because the natural state of motion is linear, and forces are needed to create a change, therefore, matter, rather than time, is the cause of changes. Causality is then attributed to material properties like mass, charge, heat, pressure, etc. In Eastern thinking, all these properties are not the causes of change; the real cause of change—which goes in cycles—is time. As a result, there are bodily cycles, perceptual cycles, ideological cycles, cyclical changes in belief systems, the rise and fall of civilizations, the circulation of money in an economy, and ultimately the cycle of birth and death—automatically caused by the nature of time. Even the universe is cyclically created, destroyed, and recreated. No force is needed to cause these changes because they occur automatically due to time. Force is only needed if we want to escape the bodily, perceptual, ideological, intentional, and value cycles. This force—which we can apply—will not change the occurrence of the cyclical changes in the world, but it can change our life. As a result, the Eastern thinking has, for the largest duration of history, not been focused on trying to change the world, make it a better place to live, or on materialistic 'progress'. It has been focused on two things— (1) accept the natural cycles and 'go with the flow of nature', and (2) try to escape this cycle. Even the efforts of changing the world only focus on escaping such cycles.

From a scientific perspective, the starkest contrast in thinking comes from the idea of causality. The Western model of the world is based on the idea that material properties and their laws (such as mass and the law of gravitation, or charge and the laws of electrodynamics) are the causes of change, and time is simply a causally inert 'parameter'. The Eastern thinking about the world is based on the idea that time is the *cause* of change, and the changes to what we perceive, conceive, judge, intend, and value, are the *effects* of time. This fundamental difference between Western and Eastern thinking leads to two radically different ideas about time. Time in the West is linear and single-tiered, but it is cyclical and hierarchical in the East. Cyclical means that changes caused by time occur repetitively. Hierarchical means that changes occur simultaneously at many levels of experience. Due to hierarchy, a larger but slower rate change embeds a smaller but faster rate of change. Therefore, there are hamster wheels inside hamster wheels, and the larger hamster wheel causes the rotation of the smaller such wheels.

The hierarchical view of reality is also 'closed' because this reality is described in terms of meanings, and all meanings are defined through relations between wholes and parts. For example, a part of a chair is called a backrest, seat, armrest, etc. only because there is a whole chair, and the parts of that chair are related to the whole. Therefore, in one dominant form, the study of reality is simply the study of the nature of the whole and the part. Thus, the world is described as an *inverted tree* in which the root is the whole, and the branches, twigs, and the leaves are successively smaller parts. Our seeing for instance is the 'whole' and the perception of color, form, and size are its parts. Then, the color is the 'whole' and red, blue, and green, are the parts. The inverted tree of all these concepts constitutes the definition of *space*. It is simply a collection of all the possibilities, that are organized hierarchically. Now, time is defined as the cycling of the experience through this closed space of possibilities—just like a hamster running in a hamster wheel.

This ideology of the world involves three ontologies—time, space, and observers. Time is identical to cyclical change, or cyclical causality, therefore, time and causality are not separate ideas. Likewise, space is identical to matter, or a closed world of possibilities. And the observer is the consciousness that iterates over the closed space of possibilities pushed by time. The observer—sometimes loosely equated to the Western idea of a 'soul'—is, however, free to leave the hamster wheel, into another world where it is not pushed by time, but truly recovers its free will to act as its chooses.

In the 'other world', change is caused by volition rather than forced by time. Therefore, if you don't want to die, you don't need to die. If you don't want to eat, then there is no hunger. If don't want to stop eating, then you can eat indefinitely. Time still appears as the changes in experience, but the changes are caused by the soul's choices, rather than due to time. In short, time is the *cause* of changes in this world, and it is the *effect* of choice in the other world. Unless we understand how time is the cause of changes in this world—which requires overturning the modern scientific ideas of change being caused by material properties—we cannot understand the nature of experience, its evolution, and the role of choices in the other world.

Thus, the ideas of space, time, matter, soul, bondage, liberation, eternity, cyclical change, mind, meaning, perception, values, intentions, etc. are so tightly interwoven, that they cannot be extricated from each other. In a few words, we cannot obtain a *complete* understanding of any topic without understanding every other topic. Conversely, if we progress in understanding one topic, then we also progress in understanding every other possible subject. This is both a boon and a bane. Those accustomed to the idea that time must be studied separately from other subjects, will naturally be baffled by the induction of other topics into the discussion. Such compartmentalization of topics is peculiar to Western thinking which divides the world into many boxes and ignores their mutual influence. In Eastern thinking, there are many boxes, organized in a hierarchy, such that the change to the lower box has a smaller effect on the higher box, but the change to the higher box affects the lower boxes. Thus, sometimes, we can use compartmentalization to simplify a subject, but that is an exception rather than the norm. The norm is that we understand everything in trying to know anything.

Time is one such topic that touches everything else, because everything evolves with time. The evolution is forced by time in this world; therefore, time is also the causal agent behind all changes, like a magic wand that seems to move everything in a magician's trick. Underlying the magic, however, is a theory of causality, which is intuitively accessible, and philosophically rigorous, but seems counterintuitive to a mind conditioned to thinking that matter moves on its own due to its own properties. There are indeed properties in matter, and matter indeed moves and changes in time. But could it be that everything that science has considered causality so far is false? Indeed, this is possible if the changes are not *in time* but *caused by time*.

This is such a radical viewpoint, that it overturns centuries of scientific thinking. Planets are not moving because of gravitational force exerted on mass; rather, they are moving cyclically due to the nature of time. Chemical reactions are not caused due to attractive and repulsive electromagnetic forces exerted by charges; rather, they are caused due to time. Weather changes, seasonal changes, and the rise and fall of civilizations is not due to material properties and associated forces; rather, they are due to time.

Except for atomic theory, which recognizes the existence of a world of possibilities, from which an observation is created, all other ideas of modern science are false in this worldview. To the extent that atomic theory describes causality in terms of physical properties like mass, charge, color, etc., it is also false. To the extent that physicists believe someday the world of possibilities will be replaced by a deterministic reality, such attempts at scientific progress are also futile. Everything based on calculus, which assumes a parameterized time, is false. And since all laws of modern science are framed in terms of differential and integral equations, which are then based on calculus and parameterized time, they are also false. Parameterized time is simply a *measure* of change, not the *cause* of change. As a result, we can distinguish between *Causal Time* and *Parameterized Time*. The former is the cause of all changes, and the latter is the measure of such changes.

A simplified view of reality is that 'space' is the domain of possibilities, and 'time' is the causal agency that selects the possibilities one after another. An 'observer' passes through a subset of such selected possibilities one after another, thereby constructing a personal trajectory. This trajectory, however, is not a *line*. It is an experienced 'world' which includes the distinction between the observer and the observed, the perceived distances to the many observed things, and a hierarchy of perceptions in the experience. The observer's trajectory is hence a relativized view of space and time, that supervenes on an absolute space of possibilities being selected by an absolute time. Therefore, different observers can experience the same world of possibilities, and hence, the world is objective (i.e. different from the observer). Likewise, since time selects these possibilities, therefore, it is also objective (i.e. different from the individual observers). What we call our 'body' is one of the many possibilities, and it is also objective (i.e. different from the individual observer). Hence, space, time, and matter are all different from the observer, and this difference leads to the notion that the study of the world—i.e. science—must be an objective description of reality, even when that description pertains to the observer's experience of the world.

In one sense, the parameterized time is objective, because it pertains to the perception by the objective body. In another sense, this objectivity is not a *public reality*, because only one observer perceives the world in this way. Therefore, the relativized space and time is objective, and yet individual. This space and time are different from the space of possibilities and the time that selects these possibilities. The space of possibilities is, for instance, partially accessible to everyone, but is not accessed by all observers.

Thus, we draw a distinction between *phenomenal time* and *causal time*. The former is relative, and the latter is absolute. Likewise, there is a *phenomenal space*, and a *causal space*; the former is relative, and the latter is absolute.

The possibilities within the causal space interact with each other because of time, and this interaction produces experiences. These interactions have different *strengths* defined by the *type* of energy exchanged—when the type is abstract, then the interaction is weak, and when the type is detailed, then the interaction is strong. They also have *powers* defined by the *amount* of energy exchanged—a faster energy exchange creates a high-power interaction, and a slower energy exchange creates a low-power interaction. The difference between strength and power is like that between *contrast* and *brightness*. Brightness pertains to the amount of energy exchanged, and contrast to the hierarchical types that produce the whole-part distinctions.

If the interaction is strong, then we see things clearly, and that clarity leads to the impression that the perceived object is 'closer' to us; such clarity would be impossible if the perceived object was 'far'. Likewise, if the interaction has more power, then we see things quickly, as opposed to if they have less power. Since distance reduces with a stronger interaction, therefore, the relative space is not equidistant for everyone—those interacting strongly with a perceived object will find it closer. Similarly, since duration reduces with more power, therefore, time is not equidurational for everyone—those interacting with the perceived object with greater power will find that they can complete the process of vision quicker. Hence, a stronger interaction reduces distance, resulting in the so-called "length contraction" in relativity. Similarly, lower power interaction increases time, resulting in the so-called "time dilation" in relativity. Hence, the relativized space and time—which we call our sense of *distance* and *duration* of things—is a side-effect of the *strength* and *power* of interaction between the possibilities.

The conceptual basis of relativity is that as brightness grows, the picture becomes hazier, and we cannot see extremely bright things both closely and clearly. To see them clearly, we must keep pictures with higher brightness farther, which is useful if we want to see the "big picture". Conversely, if the pictures must be seen closely—i.e. the "details" are important—then we must reduce the brightness, and the vision of the details results in greater contrast in the picture. The principle involved in relativity is that we cannot see the big picture and the details *simultaneously.* To see the big picture, we must decrease the proximity, and increase the brightness. And to see the details, we must increase the proximity, and decrease the brightness.

If an image is far, then the interaction must have lesser strength and greater power. If an image is close, then the interaction must have greater strength and lesser power. We can formulate a scientific principle: a stronger interaction must be slower, and a weaker interaction must be faster.

This is a principle of perception, and not a physical principle. There are no physical reasons why brighter things also cannot be close, and no physical reason why darker things cannot be seen from far. As a result, relativity is not understood when we think of the world physically, but it becomes intuitive when we think of measurement in terms of our perception.

Our inability to simultaneously know both the big picture and the details, entails that every perception involves a *sacrifice* of either the whole or the parts, and the sacrifice creates many different perspectives on the same reality, which reveal different "views" about the same world. Clearly, since each perspective is sacrificing something, therefore, no two perspectives can claim to be *equivalent*, even though they are perceiving the same world. The principle of relativity is that we have the freedom to observe the world in different ways, but each such observation involves a choice and a compromise. The choice seems easy, but the compromise produces a science.

The interrelation between strength and power, or contrast and brightness, or distance and duration, is the basis upon which the relativized space and time is called a "four-dimensional spacetime", eschewing the absoluteness of space and time, and their distinguishability. But absolute space and time need not be rejected when relative space and time are adopted; this is possible, if we can *explain* how the stronger-slower and weaker-faster interactions in the absolute space and time create the experience of relative space and time with the properties of length contraction and time dilation.

An application of the principle of relativity is that in the limiting case of the interaction becoming infinitely strong (i.e. the observer being at zero distance to the observed), the brightness must reduce to zero (i.e. time taken to see something becomes infinite). Now, two objects are infinitely close, but they are also invisible, because despite the proximity, the brightness is zero. Conversely, in the limiting case of the interaction being infinitely weak (i.e. the observer being at infinite distance to the observed), the brightness must become infinite (i.e. time taken to see something becomes zero). Now, two objects are infinitely bright, but no details can be observed, and therefore, despite the brightness, nothing is visible as the contrast is zero.

Upon this basis, we can describe the origin of the world in a new way: the world always exists as a primordial possibility, but nothing is visible if the possibility interaction strength is infinite, because despite zero distance, the brightness is also zero. The world becomes visible as the interaction weakens, the contrast weakens, but brightness increases, and the perception of many individual things is created. The weakening of the coupling between the possibilities can be attributed to causal time. Conversely, as the strength of the interactions decreases, the brightness also must go up, and the world disappears from our vision, because despite infinite brightness, the contrast is zero. The decreased interaction between the possibilities can also be attributed to causal time. These two scenarios constitute 'infinite darkness' and 'infinite brightness'. They are also the limiting conditions of our vision: we cannot see without some brightness and without some contrast.

These two limits of the world constitute the "cold death" and "heat death" scenarios of the universe. However, these are only *phenomena*. When the world is visible, it has not 'appeared', and when it is invisible, it has not 'disappeared'. The total matter or energy is always conserved (but I define this energy or matter in terms of meanings, not physical energy). But due to the effect of time, this energy can become visible or invisible. When it becomes invisible, it can be due to infinite brightness or contrast. Likewise, when it is visible, either brightness of contrast is traded off, which means that we are either seeing details without the big picture or vice versa.

The so-called "big-bang" and "big-crunch" are now phenomena of apparent or phenomenal spacetime. During big bang, the interaction between the possibilities begins to weaken, and the things *seem* to move apart, and the universe gets brighter although more obscure, creating the illusion of an expanding universe. During big crunch, the interaction between the possibilities beings to strengthen, and things *seem* to come closer, and the universe seems to get clearer although dimmer, creating the illusion of a collapsing universe. Factually, the universe is precisely what it always has been, and will always be. The seeming proximity and distance is only an effect of causal time. The world is therefore not expanding or contracting; but it *seems* to expand and contract because of time. Accordingly, what modern science calls the redshift of galaxies (attributed to the so-called expansion of the universe), is just the weakening of the inter-action between possibilities, which create the appearance of expansion, without a real expansion.

We must now draw a distinction between *change* and *motion*. The *change* pertains to the observer successively attaching itself to different bodies, considering itself that body. The *motion* pertains to a given body strongly or weakly interacting with other bodies or objects. With changing bodies, the *origin* of the space-time moves from one body to another and with moving bodies, the distances and durations of these bodies evolve. If the observer considers himself stationary—as is assumed in relativity theory—then only the motion is detected. But factually through such interactions, the observer is also getting older, and more likely, stupider. To be wiser, one must see the same world through the lens of both change and motion. Motion is not the only phenomenal reality; there are also bodily changes in progress.

In general, every change of body is caused by an interaction, and hence involves motion. Likewise, every motion is caused by some interaction with some object. Therefore, motion leads to change, and change leads to motion, and these are so intertwined that we don't understand the difference. I will discuss later in the book how the next body exists as a *destination* and the motion is the *trajectory* used to attain the body. Changes and motions are therefore as different as *goals* and *trajectories* to reach the goal. Obviously, such a separation requires the discreteness of the changes and the motions, and the situation can be compared to a frog hopping from one pond to another, such that both ponds and the hops between the ponds are discrete. This fact allows us to reconcile the modern discrepancies between atomic theory (where space and time are discrete) and relativity theory (where space and time are continuous). The reconciliation is that the effects of length contraction and time dilation can be understood from another view about space and time, that is consistent with the idea that the world is a possibility, the selection of those possibilities, the interactions between them, and a conscious observer being involved in the possibility interactions.

Philosophy is useful if it leads to new scientific ideas, which are an improvement upon current science. Philosophy is useful if it clarifies problems to a point that the solution becomes obvious. Philosophy is useless if it doesn't improve our understanding, and instead of clarifying the problem, it muddles it further to a point where the solution seems less obvious.

I am a patron of the philosophy of the former kind, and I hope to show its good side through this book. But care must be exercised in not ignoring the important problems, and not leaving problems unsolved, after they have been considered. This necessitates a journey through problems and solutions, carefully balancing the anguish to identify all problems with the anxiety to quickly solve them. I have tried to make the novelty easier by going from what we easily accept to be true, to what follows from this acceptance, gradually. The access to this worldview is possible through a stairway of ideas that progress gradually. But this stairway includes some unfamiliar steps. Since the topic of time cannot be extricated from other topics, therefore, this not a book *just* about time. But since this *is* a book about time, therefore, the discussion always returns to time after journeying through other topics. When causality is attributed to matter, then each type of behavior requires a different particle, property, force, and law, but if causality is attribute to time, then each phenomena requires us to think of a different hierarchy of the time cycles acting on a different hierarchy of wholes and parts, and creating different strengths of interactions between them. Conceptually, the latter approach is also simple, extensible, and complete.

One final note to the reader, before we dive into the specifics. The problems that this book deals with need a reconceptualization of reality. This means (a) thinking of the world in terms of choices and possibilities, (b) instituting a central role for observers that have past memories and future goals, (c) understanding how choices are driven both by memories and goals, apart from the present circumstances, (d) that since memories and goals exist as meanings, therefore, even the present reality must be described as meanings, (e) that the possibilities that choices select from must also be meanings, and (f) through such reconceptualization of reality, overcoming the mind-body paradoxes, and limited ideas about matter and causality.

The first four chapters of the book discuss the unsolved problems of classical mechanics. As we go through these problems, I will also discuss the reconceptualization of space, time, and the observer, that solve these problems. Each chapter discusses a specific question around the nature of time, the potential solutions, and their respective problems. Once we have understood why reality must be viewed differently, then we are equipped to interpret physical theories such as relativity, quantum mechanics, and thermodynamics, diagnose their problems and understand the solution.

The last four chapters are devoted to the understanding of the problems in special relativity, quantum mechanics, thermodynamics, and cosmology, but without the foregoing first four chapters, this would not be possible. The first four chapters are more philosophical, while the last four chapters are more scientific. I understand that modern science doesn't work this way, but its problems have remained unsolved for a long time by using the conventional methods. Therefore, an alternative approach can help.

This book is written in a question-and-answer style. Each chapter poses a well-known question about time, discusses the well-known positions and answers to this question, discusses the flaws in these answers, and then proposes an answer. The questions are sequenced from simple to complex, and therefore, the successive chapters build upon the previous answers. Thus, every successive question and answer uses the answers of previous chapters. By the time we get to the end, we would have touched many topics, but due to the progression, we will also see a simple answer to all the topics.

1
Does Time Pass?

April, come she will
When streams are ripe and swelled with rain
May, she will stay
Resting in my arms again
June, she'll change her tune
In restless walks, she'll prowl the night
July, she will fly
And give no warning to her flight
August, die she must
The autumn winds blow chilly and cold
September, I remember
A love once new has now grown old
— Paul Simon

Waking, Dreaming, and Deep Sleep

During our *waking* experience, time seems to move from the past to the present to the future, and therefore, we tend to believe that time passes. The problem is that if the past has ceased to exist, and the future hasn't yet arrived, then how are we able to *think* of the past and the future? You might say: We have memory of the past, and our minds can combine ideas that they have previously acquired, using our capacities of reasoning. This argument draws upon a *computational* idea of the mind, where a computer stores data as the memory of the past, and logically derives their conclusion using reasoning. It draws a wedge between the senses and the mind—namely, that the mind is not perceiving a reality; only the senses are capable of perception. That there is no mental intuition; there is only sense perception.

This is when it becomes important to talk about the *dreaming* experience. During dreams we can see things that we have never seen before. Even those things that we have seen before are combined in ways that aren't rational. So, because we see things that we haven't seen before, we cannot say that we are recalling the memory of the past. And because we can combine what we have seen before in irrational ways, therefore, we cannot say that new thoughts arise due to rationality. In short, both contentions of the computational idea of the mind are proven false during dream experiences. To overcome this problem, we must say that the mind is not just a record of the past, and a capacity for rational thought, but is also a 'sense' capable of perception[1]. But if the mind is perceiving something, that did not exist in our memory, then where is it perceiving it from? Of course, such perception is not unique to dreaming, because it exists during waking as well, when we are involved in irrational, imaginative thinking. At this time, we seem to access a world that we may not have seen by our senses during waking.

Mental intuition underlies most of the scientific discoveries because most scientists will tell you that they 'saw' a solution completely and suddenly. Subsequently, rational arguments for this solution are presented, which makes us think that science is about rationality, when it mostly depends on intuition. For example, Ramanujan saw mathematical theorems in his mind, and even though he could not prove them, they turned out to be true. Newton saw the laws of motion in his mind, which then became public knowledge later. Some science fiction writers have the seen the possibility of airplanes and telecommunication in their heads before they became public knowledge. Again, if we have such intuition, where are we getting it from? Shouldn't there be a real world of ideas from which the mind intuits the ideas? How else can we explain the suddenness of the discoveries? But if these ideas are intuited from a preexisting world of ideas, which then become available to everyone through a rational expression of the ideas, then the ideas must exist even when we are not perceiving them sensually. However, this goes against the idea that the world is changing, and the past and the future have ceased to exist. Mental intuition challenges the conclusions arrived at from the straightforward sensual perception of the world.

The problem gets harder if we consider *deep sleep*, when we aren't sensually perceiving or mentally intuiting the world, and there is no sense of past, present or future. Upon waking up, we realize that the world around has changed, so, the world must be objectively real in the sense that it changed even when I was unaware of it. Then, we also say that there was a time when I was awake, and we can imagine a future when we would be dreaming or in deep sleep again. Deep sleep basically detaches our conscious sense of past, present, and future, from an observer-independent past, present, and future. Upon this basis, we say that the world must be changing irrespective of my awareness of change, and therefore, the world must be studied *as if* I did not exist or I was not aware of the world.

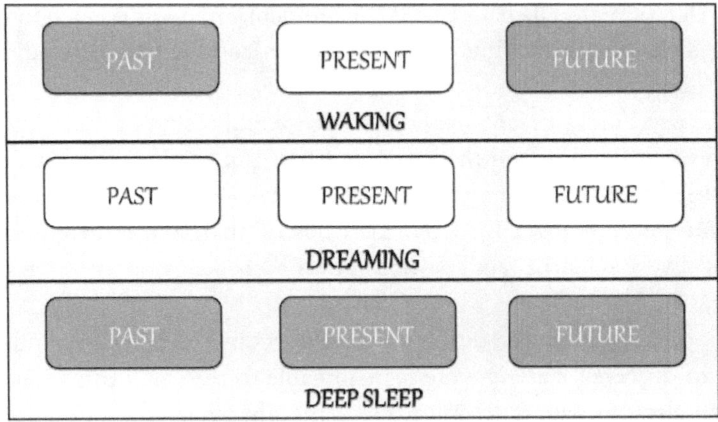

Figure-1 The Three States of Conscious Experience

Thus, we have three ideas about time based on experience: (1) during waking, time passes and the past ceases to exist, (2) during dreaming or imagining, the past and the future also exist, although by comparing it to waking, we distinguish the past and the future from the present, and (3) during deep sleep, there is no past, present, or future, not even an experience of time passing, but upon waking, we do recognize the passage of time. In the first case, time passes, and past and future don't exist at the present. In the second case, past and future exist at the present, but by comparison to waking, we do speak about passing. In the third case, there is no past, present, or future, but by comparison to waking, we speak about passing. How do we reconcile the three contradictory ideas about experienced time?

The dominant scientific response[2] to this problem is to derecognize dreaming and deep sleep, and only consider waking. That makes us believe that the world is changing, and the past and the future don't exist at the present. Of course, even while waking, we do use our memory of the past, and the goals for the future, but because this memory is conceptual, rather than physical, science also takes the next step of derecognizing the mind's effect on the world, because that fits well the dominant scientific dogmas around materialism. As a result, the purpose of a scientific theory is to describe the sequence of events in an objective, material world. The mind-body separation delinks the problem of experience from that of material change, and the problems of memory, imagination, goal formation, and their effects on our experience, are all deferred to the psychologist.

The psychologist recognizes the existence and reality of the mind, but he says that the past and the future only exist in our 'heads'[3]. But if you can see the future, that nobody else has seen yet, and that future will become public knowledge later, then how is that future merely in our heads? Every scientist sees some future in their heads before it becomes reality.

So, the claim that the future only exists in our heads is naïve and contrary to observed facts. We are factually able to see the future, and then we can also translate that vision into something that others can also see. Therefore, what we are seeing in our thoughts, dreams, visions, and creativity is not always in our 'heads'. It is rather something objective which is accessible to everyone but may not be accessed by everyone. A process of mental intuition is involved in accessing this reality, which requires us to say that the mind is also sense, but it perceives what is not always sense perceivable.

Mental intuitions, of course, work with waking experience, but in more subtle ways. For example, we can intuit other people's pains, due to what is called 'empathy'. We can sometimes know what others are thinking, or if they are lying, if they have evil intentions, or what they are planning to do. None of these things are sense perceivable, but they are mentally intuited. This ability to perceive what apparently lies in our 'private spaces' even while we are awake, requires us to say that the mental intuition exists always. Through this intuition we give meaning to our world, which is always involved in the act of interpreting the sense perceptions. What is meaning? It is the act of attaching concepts to percepts, of gleaning the purpose underlying things, of understanding if someone believes what they are saying, or the person's nature, such as their likes and dislikes, moral values, etc.

When the role of the mind is broadened to include such intuitions, then it becomes easier to talk about the nature of reality as comprising of two distinct *aspects*—a sensual aspect which is perceived through the senses, and a conceptual aspect which is gleaned through the mind. The world must now be described as *symbols* of meaning that combine the sensual and the conceptual. These two are not separate 'substances' as was assumed in the Cartesian mind-body divide, which underwrote the initial developments in modern science, and subsequently came to be accepted as the doctrine of all reality. Indeed, mind and body are not even exclusive to living beings such as humans. Rather, there is an objective conceptual reality—which we can call 'meaning', that exists in inanimate objects such as books, music, and art, quite like the physical properties perceived by our senses are objectively present in inanimate objects. The capacity for perceiving these qualities is unique to observers, such as humans, so just as our sense perception is decoupled into a 'sense' and an 'object', similarly, our mental intuition should be decoupled into a 'meaning' and a 'mind'. The theorems of mathematics, the theories of physics, the forms underlying art and music, and the stories depicted in literature are varieties of meanings that exist objectively in science, art, music, and literature—when they are embodied into sensual artifacts. But to explain the process of mental intuition, we must say that these exist as meanings even when they are not perceivable via the senses.

This hypothesis—which I call the *semantic worldview*—helps us reconcile the disparities between waking, dreaming, and deep sleep experiences. In this hypothesis, dreams are visions of a conceptual reality, that has no physical embodiment. Scientific, literary, and artistic creativity also taps into this world of meanings to produce objects that don't just have to be seen, but also *understood*. This reality exists in our minds while we are awake, and imagining, planning, or thinking about the past or the future. And this reality in our minds can be perceived by others—if they have mental intuition. Therefore, the waking experience accesses a conceptual world, although it could exist in the past, present, or future. The dreaming experience, similarly, accesses the same conceptual world, in the past, present, or future. The basic difference between dreams and waking is simply that our dream experience is private, while the waking experiences can also be public. During deep sleep, however, both sensual and mental perceptions cease, but that cessation neither indicates an end to this world, nor to changes to it.

Measurements vs. Observations

The semantic worldview is crucial to the rest of this book, and it stems from a broader appreciation of our observation, rather than the *measurement* of physical properties upon which science rests. The instruments also have concepts—like books, art, or music—but they don't have *senses* and *mind* to perceive the world. The absence of the senses and mind entail that a measurement will show the *effects* of the presence of sensual and conceptual realities, but the measurement cannot know the *causes* of those effects. Thus, the existence of meaning in inanimate objects must produce observable effects which cannot be explained using the assumptions of a mind-body divided world. The inability to predict and explain these observations will then lead to the incompleteness of scientific theories, but we would never be able to surmount these gaps in theories unless we go back to the idea that we are not merely perceiving the world by our senses, but also via the mind.

If the reality of mental intuitions is recognized, then we can recognize the reality of meanings, and incorporate the causal effects of meaning into a scientific theory. If we keep insisting on physical measurements, then we will *think* that there is nothing other than physical properties, but our theories using these properties will be incomplete. On the other hand, if mental intuitions are recognized, and the concepts intuited by the mind are given causality, then even the results of interactions between inanimate objects—such as that between a measuring instrument and its measured object—would be explained using perceived properties and intuited concepts.

However, before we get too excited about these proposals, we must ponder the mind-body divide again. What is the body really? Is it physical properties, such as mass, energy, momentum, charge, etc.? Or are such properties the artifacts of modern science, even though every observation uses sense perceptions like taste, touch, smell, sound, and sight? I ask this question because the mind-divide creates problems only if the world is physical properties. If we describe the world as sense perceptions—rather than physical properties—then both our percepts and concepts are simply two different types of *ideas*. The sense perceptions of color, form, taste, or smell are no less conceptual than the concepts in the mind. For example, we habitually use color shades, nouns about shapes, tastes, and smells, which are *qualitative* rather than *quantitative*. We perceive the world as these qualities, but modern science objectifies this qualitative world into quantities by selecting a standard that objectifies some quality into the unit of a property. Mathematical equations now compute quantities, but the meanings of the words such as 'mass', 'energy', 'momentum', or 'charge' is still given through sense perception along with an instrument that embodies the qualities. Such computations and measurements are not the real issue. The real issue is that we start believing that the world is quantities or numbers, rather than the qualities, because this worldview contradicts the qualities in the mind.

To understand this problem, let's use an illustration. Let's suppose that we define a property called 'sandwich', and denote it by a letter S, by picking a two-slice sandwich as the standard. Every other sandwich can now be compared to this standard, and we can say that some large sandwich is 3S. Our mathematical equations can now use the property of S, and compute numbers. But all such numbers must ultimately be some multiple of S, and we don't understand S itself as a number; we know it as a 'sandwich'.

At the dawn of modern science, this problem was recognized, and a distinction between *primary* and *secondary* properties was drawn. The primary properties were mass, energy, momentum, length, time, etc. while the secondary properties were color, taste, smell, shape, etc. There were two reasons offered in favor of the primary properties. First, it was said, that science must describe a world independent of our observation, but this idea was ultimately false because we never measure these physical properties; we only observe the sensations. Second, it was supposed without adequate justification, that the primary properties are far fewer than the sense perceptions. This seemed to work in the limited scope in which Newton's physics was formulated, because the physical properties then were indeed few. Obviously, if our physical properties exceeded the sense percepts, then we would be better off using sense percepts. The only reason why we would prefer physical properties if the sense perceptions were reduced to physical properties, and reduction assumes that physical properties are fewer.

This assumption of classical physics is now known to be false. For example, atomic theory employs mass, charge, momentum, energy, and angular momentum, as did classical physics. But atomic theory adds to this list other properties like spin, isospin, color, strangeness, along with dozens of unique combinations of these properties, which are now known as *fundamental particles*. Each such particle also has a unique value of these properties, so in addition to these properties, we also have dozens of natural constants which must also be treated as fundamental constants of nature. When we combine all the types of properties, the numerous particles, and the unique values of the natural constants, the world of primary properties is on longer simpler as compared to the world of secondary properties. Incidentally, since such properties must be measured using everyday instruments—e.g. the Large Hadron Collider—which requires enormous amounts of equipment, the complexity of the primary properties adds to the secondary properties, rather than substituting them completely. Now, the properties meant to explain the basic sense percepts far exceed the basic perceptions!

Einstein had an example to illustrate the use of physical properties: he compared these properties to tokens you get in exchange for a coat at the entrance of a ballroom. When you are finished dancing, you come out and return the token to get the coat back. This process works well if the token is smaller and lighter than the coat. If the number of tokens you must carry in exchange for the coat are bigger than the coat, then you might as well carry the coat, instead of these tokens! After all, we don't really care about those tokens; what really matters is that we get our coat eventually.

Classical physics grew out of the idea that the coats were complex, and the tokens were simple: our sense perceptions were complex, and the physical properties were simple. This is no longer true. Add to this fact that we ultimately only need the coats, and not the tokens. When the burden of carrying the tokens exceeds that of carrying the coat, then we should ideally discard the tokens. In physical theories, this means we must now consider the qualities of perception themselves as causal properties, because this approach to theory formation is simpler relative to the current approach.

This approach has the added advantage that it immediately resolves the mind-body problem, because both sensations and concepts are ideas. Some of these ideas are associated to our senses, and some of them are associated with our minds. Of course, we will now have to make the senses and the mind as the instruments of observation, quite like we presently treat a meter, a clock, and a kilogram as the instrument of measurement. This is a surprising U-turn from the perspective of classical physics, but it is not only more parsimonious and simpler from the perspective of modern physics, but also the preferred mode that should be used in describing our experiences.

Possibility and Reality

Once we accept the semantic worldview, then we can return to the problem of the perception of meanings, which, as we have noted, are not accessible to everyone. Some people have intuitive minds, and they can perceive the meanings better. Others have sharper senses, and they can perceive the sensations better. In short, sensual and mental acuity is important to know the world correctly, and we are no longer talking about a universal standard for all properties. Since we can choose not to look at the world, and thereby not know it, therefore, the world of meanings must also be treated as a *possibility*[4]. This possibility can be accessed if (1) we try to access it, and (2) we have the perceptual and intuitive capacity for it. Without the attempt and the capacity to know, the world is an objective possibility. This idea about objectivity is commonly accepted in everyday life. For example, you can be cured of a disease *if* you take a medicine. You can acquire knowledge *if* you read a book. You can satisfy your hunger *if* you eat the food. The medicine, the food, or the book do not forcibly act upon us.

While dreaming, thinking, imagining, or creating, we access these possibilities by our mind, and while waking, we access them by our senses. For example, we might not sometimes see what lies in plain sight, or we might see it, but change our focus, like we avoid thinking about the bad past or a painful future. Such access to possibilities disappears during deep sleep, but the objective changes to possibilities remain. The deep sleep experience tells us that the world changes even when we are completely unaware of it. So, the world must be *objectively* present. Since in dreaming we can become aware of it, therefore, this objective reality must be *possibility*. Finally, because this possibility is converted into a perception, hence, the possibilities must be divided into those that are mentally and sensually accessible, respectively.

The trouble is that the scientific notions of objectivity are constructed to ensure that everyone can see everything all the time[5]. According to science, you cannot choose what you want to see; you cannot change your focus from one thing to another, because everything always influences you. The problem of time's passing is therefore rooted in the problems of scientific objectivity, and its solution requires changes to this objectivity. Objectivity is not a thing that always acts upon everything everywhere. It *can* act but it doesn't; it exists as a possibility that must be made to interact with other things to produce an effect, but those things are possibilities too.

Time and Eternity

The notion that the world exists as a possibility makes reality *eternal*. Even when Pythagoras's theorem was unknown, Pythagoras did not *invent* it. He only *discovered* the eternal possibility. This idea is like Greek Platonism[6], in the sense that there is an eternal world of forms, which become accessible to us at different times. However, this idea is also different from Platonism in two ways. First, the Platonic world comprises *ideal forms*, but the reality that we discover isn't always ideal. Second, the Platonic world is also beyond this world, but the reality we discover is this world, although not always perceivable. The eternality of possibilities and the temporality of this world are thus not contradictory, and everything that has existed in the past, or will exist in the future, is also present now—although, as a possibility.

Time is therefore the *mechanism* by which the possibilities are discovered. This discovery involves a few different notions. First, some possibilities are *more* possible than others at any time. For example, at present, forests and gardens are being replaced by cities and buildings, therefore, the possibility of observing forests and gardens has reduced. Likewise, dinosaurs cannot be observed at present, although they existed in the past, and could exist in the future. The reduced possibility of observing certain things, however, doesn't end the possibility of our *imagining* their nature. We can still access the possibility world through our minds, and in that sense, the past, present, and future exist always, even if they are not perceivable. Second, to access these possibilities through the mind or the senses, our consciousness must be *drawn toward* these things. Our consciousness too is a possibility that can perceive and conceive the world and can be focused or withdrawn from these things. Time affects our consciousness too, by directing it toward different kinds of possibilities—either through the mind or the senses. Why do some ideas become collectively popular at some times, and unpopular at other times? Why is the world driven toward consumption of certain types of commodities at one time vs. another? The popularity and unpopularity of ideas and things means that time also affects our consciousness, and this influence can be individual or collective. Therefore, when we speak about time's effect on possibilities, we must recognize two such kinds of effects. First, certain things become likelier than others, and hence we are drawn toward them. Second, certain ideas become more popular, and the demand for these ideas then leads to the popularity of things representing them.

The Role of Free Will

Now, I'm not suggesting that time *determines* our perception—which would be a denial of free will. For example, it is entirely possible that Pythagoras could have just worked out his theorem in private, thought of it in the mind, and not have put it on paper, or never shared that proof with others. This complicates the problem of possibility and its selection by time; specifically, we must think of the emergence of a world as not being contrary to the idea of free will. This reconciliation involves the idea that we have an acquired *nature* or *personality* of habits, likes, etc. For example, if you develop a habit of waking up early, then you will automatically be woken up early by time. If you develop a habit of eating on time, then at that appropriate time, hunger will automatically arise. If you develop an aggressive nature, then every situation would automatically lead to the feelings of aggression. This seems like determinism, but only because we don't realize that we can change our nature, habits, personality, likes and dislikes. Our choice is affected by our ability to reject the automatically arising desires, especially if they are contrary to our goals. Even if our consciousness is pushed toward a possibility, we can resist that push, which changes our nature such that we don't even feel the push. Therefore, time acts upon us by waking us up, causing hunger, creating feeling of aggression, hurt, pain, or lust. But time is acting on our acquired *nature* rather than on *us*. Our ability to resist our nature requires us to draw a distinction between a *person* and their *personality*. If this distinction is not drawn, then time's action leads to determinism.

If the push of time is accepted by us, then we become aware of a possibility. If the push is rejected, then we become unaware of the possibility. Time's effect on consciousness, thus, need not entail determinism; it could just *seem* deterministic if we always *accept* rather than *reject* time's push.

There are other ways in which time's push doesn't lead to determinism. For example, even if time pushes us toward a certain goal, there are still alternative methods of attaining the goal. A longer route to our goal may be moral, and the shorter route may be immoral. We have the choice to determine not just the goals, but also the methods by which we attain them. Hence, our personality is not just *what* we like and dislike, but also *how* we achieve these goals. An honest person will, for example, try to fulfill their desires honestly, but a dishonest person will adopt dishonest means to fulfill the same desires. Thus, determinism can be violated in two ways— (1) we can change our goals, and (2) change the methods of achieving them.

Employing different methods to attain a goal also often requires a different kind of skill. This leads to another kind of choice—namely, that we can spend the time acquiring a new skill, upgrading an existing skill, or just using the skill we have. Similarly, our abilities do not just determine *how* we do things, but also *what* we will do. For example, if you are incapable of being a sportsperson, then you might abandon the goal of being a sportsperson and pursue the goals for which you already possess the abilities.

Thus, upon closer inspection, we end up with a bidirectional causation in which time creates goals, which we can accept and reject. And our abilities and skills restrict the possible goals that we are going to pursue. Therefore, what we accept or reject depends on two kinds of natures—(1) what we like and dislike, and (2) what we are capable or incapable of. Now, what we called 'possibility' above, can be nuanced to also denote abilities and skills. Matter is now like a tool in the hands of an operator. If you don't have the tools to build a house, you either try to acquire the tools, or abandon the goal of building a house. Our choices are caught between two constraints—what we *want* to do, and what we *can* do. We can change our wants by changing our persona, and we can change our skills by acquiring new skills.

Our goals determine *what* we will do. Our skills decide *how* we will do. And our free will determines *who* will do what and how. Determinism arises only when the questions of what, how, and who are collapsed into a single physical entity governed by a mathematical law. This identity of what, how, and who, then leads to the idea that time must also be physical and cannot be subjective. All such ideas—which pervade modern science—are false, and when the foundation is flawed, then the superstructure is prone to mistakes. The correct model of causation is that which eschews determinism because it is incompatible with the idea of choices creating a lived time.

Causality in Possibility

We cannot take away causality from possibility because possibility constrains choices. If time is the agency that selects a possibility from a set of possibilities, this doesn't make the possibility 'inert'. The non-inertness is obvious in the case of our consciousness because we suppose that we have free will. But there is a deeper sense in which the possibility can influence the choice—if the possibility is reduced, then the choices are also reduced, and if the possibility expands, then the choices also expand. Thus, by the expansion and contraction of possibilities, time's effect can be modified.

The causality in possibility arises because choices have *consequences* and we can distinguish between *good* and *bad* choices. What is good? It is that which expands the possibilities, choices, freedom, and free will.

What is bad? It is that which contracts the possibilities, choices, freedom, and free will. When possibilities expand, then we choose from a broader set of alternatives. When possibilities contract, then we choose from a smaller set of alternatives. Thus, in one form, possibility has causal effects if choices have consequences, and the possibility expands or contracts as the consequence of choice—it pushes choice toward the 'good' instead of the 'bad'.

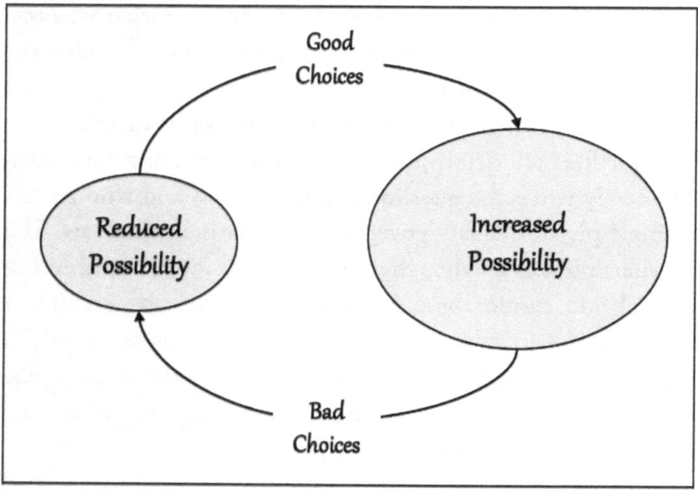

Figure-2 The Consequences of Choices

In another form, possibility will have causality because any choice alters the world, and makes some earlier possibilities impossible, and some previously impossible things become possible. The possibility may be eternally conceivable, but it is not eternally *perceivable*. This is easily explained by the fact that dinosaurs don't exist at present, although in principle they are conceivable. Eternal possibility is thus different from what is possible right now. If choices alter the current possibilities, then they cannot be inert. We can call the conversion of previously unperceivable reality into something that can be perceived as the *effect* of a choice upon a preexisting possibility.

The possibility is not inert as it has two kinds of causal influences—*effects* and *consequences*. As we make choices, some possibilities become feasible while others become infeasible. I call this the *effect* of choices. Likewise, some choices expand the possibilities, while others reduce those possibilities. We can call this the *consequence* of choices. And the effect and consequence can be distinguished from the *result* which is that an immediate outcome is produced by selecting a possibility through a choice. The result arising from a choice is generally attributed to the choice, and not to the possibility, and due to this, we assume that possibility has no causality. But if possibilities evolve and expand/contract with choices, then choice is the driver of change (i.e. if no choice were made then possibilities will not evolve or expand/contract) but that doesn't make the possibility inert.

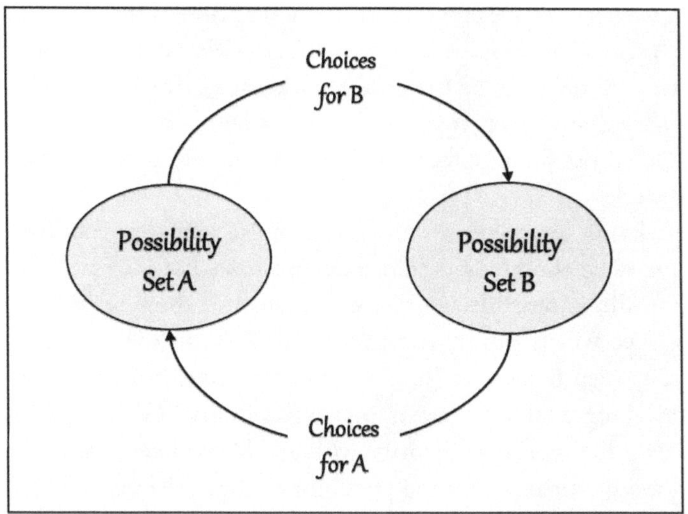

Figure-3 The Effects of Choices

In that sense, we can say that the choice is the *dominant* cause of change, while the possibility is the *subordinate* cause, however, both causes are responsible for effects. The causality arising out of the postulate of a domain of possibilities selected by choices, is hence quite different than deterministic causality which only yields *results*. Instead of being 'inert' or 'inactive', the causality in possibility is broader than the causation of determinism.

Possibility vs. Probability

The causal power in possibility is obscured when possibility is reduced to *probability*. Probabilities are not causal because a 50% likelihood of a coin toss turning up heads depends on the coin being tossed. The probability simply says that *if* the coin is tossed, then 50% of the outcomes will be heads. It doesn't say whether, where, when, how, and why the coin will be tossed. Likewise, if the coin is tossed, it could turn up heads for the first thousand years, and then tails for the next thousand years, and the probability of heads and tails would still be 50%. Probability is a measure of the *effects* and has little to do with the *causes*. Possibility, on the other hand, is causal—it decides what *can* and *cannot* happen. Thus, when we say that time causes change, we don't mean to say that time is the *only* cause, and matter or consciousness are 'inert' or acausal. We mean something more sophisticated: time can be the *predominant* cause. This is a novel concept of causality and it overturns two key ideas about causality: (1) there is always one cause for an effect, and (2) this causality is accepted only if it is irrevocable.

The relation between possibility and choice is that possibility decides what *cannot* be chosen, and choice determines what *is* chosen. In a simple sense, the impossible cannot be chosen, and the possible cannot be denied. Free will is simply the ability to constrain the possibility. This is typically seen if we withdraw our consciousness from the world; this withdrawal represents the constraint on possibility—we are not prepared to perceive the world or identify with it. We will see later, while discussing atomic theory, that the possibilities about the world evolve with time. The evolution of the possibilities—which represents the expansion or contraction of possibility—therefore must be distinguished from the selection of a possibility. The evolution determines what is *impossible*, and choice determines what is a *fact*. Between the fact and the impossible lies the non-chosen possibility.

Thus, we arrive at a sophisticated understanding of causality in which there is a 'space' of possibilities from which time selects one possibility to create a future. The 'space' however is not infinite or inert. Rather, this space can also expand and contract, and the *form* adopted by the space also constitutes a *choice*. But this choice is different from the choice in time.

In this view of the world, there are three main ontologies—space, time, and observers—which are redefined to address the three problems of passing time: (1) everything is not possible all the time, (2) everything is not visible all the time, and (3) everyone doesn't have access to everything. The notions of everything, every time, and everyone, necessitate the three ontologies. We can accept that this viewpoint is speculative presently, and I will refine and develop this view over the course of this book, demonstrating how it addresses all the problems of our experience and the reality underlying it. For now, let's assume this view and see where it can take us.

Subjective and Objective Time

These three ontologies also lead us to three kinds of time: (1) a personal time in which we make choices of expanding and contracting our consciousness, thereby accepting and rejecting experiences, (2) the objective existence of possibilities and impossibilities which change over time. The latter time can be called *objective*[7] while the former time must be considered *subjective*. There is a sense in which the world exists independent of our observation. If I close my eyes, then the world doesn't disappear. However, this objective reality must be further divided into that which can be eternally intuited mentally, and that which can also be observed sensually. The eternally intuited possibilities constitute an objective space, whereas the temporally observable sensual reality constitute a subset of that space selected by time. The space and time are distinct from the observer because the observer can disengage his consciousness from the worldly perceptions and intuitions.

Now, time has a causal role in both subjective and objective variants. This causality is also associated with *choice*. For the subjective time, we choose to engage with a possibility and thus create our personal experiences. Meditative practices teach us how to withdraw our consciousness from the world[8], which tells us that engagement with the world is a choice. For the objective time, again a choice is involved because all that is eternally possible is not visible at a given moment. Therefore, some of the eternal possibilities must be selected temporally to create a sensually observable reality.

Thus, we can distinguish between an objective and subjective time, and the objective time comprises two things: (1) that which everyone can potentially perceive by the senses, and (2) that which some people can potentially conceive with their minds. These two objectivities can be called 'perceivable' and 'conceivable' in the commonsensical use of the words where we accept that what doesn't exist right now is also possible. Whatever is perceivable must arise from what was previously conceivable, so all that is conceivable is the domain of 'possible'. But, to the extent that all that is possible may not be conceived by anyone (at the present), we must distinguish between what we are conceiving and what is possible. The possible is also distinct from what is perceivable, and then, from what is perceivable to an individual. Thus, we arrive at four distinct sets shown in Figure-4.

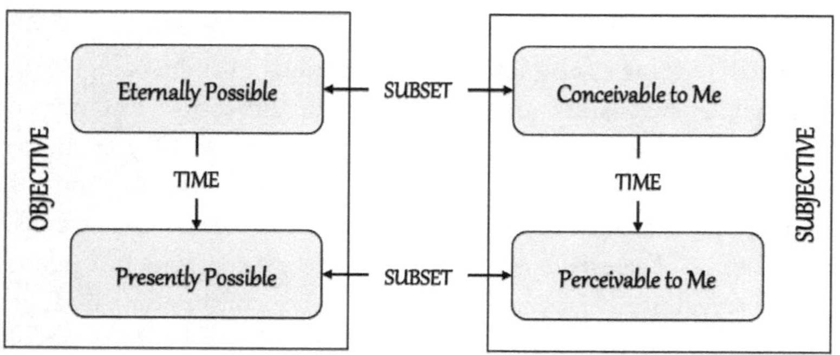

Figure-4 Subjective and Objective Time Differences

Since the objective time involves a choice, therefore, objectivity is also due to a subject, although this subject must be the cause of changes that are perceivable possibilities for everyone. However, since all perceivable possibilities for all other observers are created by time, therefore, this subject constitutes a Universal Observer[9]. The need for this observer will become clearer in the next chapter after we discuss the mechanisms of time's passing and why the failure of physical notions of time's passing necessitate a Self-Moved Mover that creates the succession of moments via choices. Thus, taken together, both subjective and objective times are due to choices.

The net result of drawing a distinction between objective and subjective times in this way is that these notions are not contradictory for two reasons. First, both kinds of times can be attributed to observers, and their choices. Second, the Universal Observer only creates a perceivable possibility, rather than determinism, so time's action doesn't contradict personal choices. Objectivity and subjectivity are therefore not mutually contradictory ideas.

Takeaways from Time's Passing

The discussion about time's passing, and the solution to the problems it poses, therefore, leads us to some key ideas that will prove useful throughout the book. We can briefly recapitulate these ideas here:

- Waking, dreaming, and deep sleep present us with three different ideas about the external world. Due to deep sleep, we know that the world is objectively real, independent of our observation. Due to dreaming, we know that there is more to the world than we observe at the present, during the waking sensation and intuition.

- The contrast between sensation and intuition leads us to the mind-body problem, which is unresolvable in a physical worldview. The classical premise of physical properties being simpler than sensed properties is now known to fail spectacularly in atomic theory.

- Given the scientific problems with physical properties, and the difficulties they pose in understanding our experiences, we can shift our thinking from a physical to a semantic view of reality. In the new picture, both sensations and concepts are ideas; they also require different kinds of observation instruments or 'senses'.

- Armed with this revised worldview, we can now revisit the problem of past and future existing in the present and draw a distinction between *possibility* and *reality*. The former is identical to what we can conceive, and the latter is identical to what we can perceive. Dreams access a possibility, which exists eternally. Therefore, the past and the future also exist at the present, although as a possibility.

- From this eternal world of possibilities, a subset of possibilities become sensually perceivable. Therefore, time has a role in converting some conceptual reality into a perceivable reality. This perceivable reality is also a possibility—but a sensually accessed possibility.

- We call these objective realities 'space' and 'time', but they are not the entities of modern science. Space is the domain of possibilities from which subsets of current possibilities emerge due to time.

- The existence of some possibility doesn't necessitate our experience of that possibility; similarly, our withdrawal from that experience doesn't entail the cessation of that possibility or reality. Therefore, there is an objectively real world outside our awareness.

2
How Does Time Pass?

The wheel is turning, and you can't slow down,
You can't let go and you can't hold on,
You can't go back, and you can't stand still,
If the thunder doesn't get you then the lightning will.
—Bill Kreutzmann

Event Ordering vs. Change

In modern science, objectivity is defined as that which everyone can always see, which is false because the mental states such as thoughts, and physical states such as that of pain are also objective and by reducing objectivity to something that everyone can observe, such mental and physical states are excluded from the study of reality. The dominant scientific position is that the goal of science is to describe matter *as if* we did not exist, even if we are involved in the experimentation. Thus, physical pain and mental thoughts are definitionally outside scientific inquiry. Reality is defined to be that which everyone must accept, without a choice, and it is equally governed by the same natural laws of logic and mathematics, and since these laws are *logically necessary* in all cases, hence, even if things occur with our awareness, they necessarily must happen even without us.

When change is objectified through mathematics in this way, then the *data* for science is a *strict ordering* of events (i.e. that the events occur *before* and *after* other events). Strict ordering is essential to form cause-effect relationships between events themselves (i.e. without an observer's conscious intervention) because the cause is before the effect. The events that occur prior must cause the events that occur later[1]. If this order between events were disrupted, then the cause-effect relationships would also be disrupted. Thus, to formulate a scientific description, we must clearly know what happens before or after. By event ordering, we can number events as first, second, third, etc. Now, there is a clear direction in which causality flows—from the first event to the second event to the third event, and so on.

The ordering of events, however, only produces the sense in which time flows in a direction—i.e. from one event to another—but it doesn't explain *how* time flows. Indeed, the collection of all events could also be treated as 'positions' along a time-axis, mimicking the ordering of positions along the space-axes. Since all positions in space can exist at once, the strict ordering of events along a time-axis doesn't tell us how time passes—i.e. What creates the *present,* the *past,* and the *future?* If the events on the time-axis were points on a line, then the changing present would correspond to a moving point on the line. We could designate this moving point as the moving origin of the time axis and say that the '0' of the axis is the present, while the negative numbers are 'yesterday' or 'day before yesterday', and the positive numbers are 'tomorrow' or 'day after tomorrow'. The sequence of events, and the continuously changing origin, are thus separate ideas. The time-axis gives us the ideas of *before* and *after,* whereas the present gives us the ideas of *past* and *future.* Even if the sequence of before and after remains unchanged, the ideas of the past and the future change continuously.

Thus, in an objective sense, the ideas of 'before' and 'after' are universally asserted, but the 'past' and 'future' cannot be universally asserted, because the 'present' is constantly changing. How does the present change?

The Purported Time-Object

One answer to this problem is to treat the change in the position on the time-axis just like the change in the position on the space-axis. The latter change arises when a particle moves on the space-axis, and to support a similar intuition about time, we would require another particle to move on the time-axis. The trouble is that unlike the space-axis, where many particles can exist simultaneously at many positions, on the time-axis only one particle can exist for all time. This is because we presume that the past and the future don't exist, and only the present exists, and many particles on the time-axis would entail many definitions of the 'present' which would then entail that even the past and the future are real for some people[2]. It will also entail that some observations are not unreal for some observers. Thus, we are led to the idea that the time-axis must have only one moving particle, which we can call the *time-object*. Just as the position of a particle on a space-axis changes as the particle moves along the spatial dimension, similarly, we could say that the motion of the time-object causes the time to pass. The problem is that the time-object is not an ordinary particle in space, because ordinary particles move *in* time, whereas the time-object *moves time*.

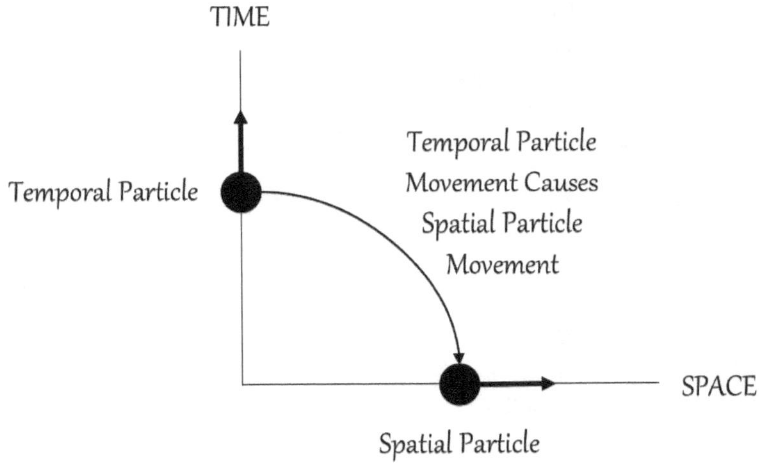

Figure-5 The Notion of a Time Particle

What is the time-object? This would be merely an interesting philosophical question if not for more pressing issues that arise from the claim that such an object exists. The pressing issue is that if we suppose that the motion of the time-object creates time for the spatial particles, then what causes the movement of the time-object? After all, the movement of the time-object also necessitates the notion of some 'time'. What is the cause of the 'time' that causes the time-object to move, which then creates time for the particles moving in space? Obviously, if the problem of time for the particles moving in space is solved by postulating an object moving on the time-axis, then the problem of the movement of the time-object must be solved by postulating yet another time-axis and another time-object, which will then require yet another axis, and time-object, *ad infinitum*. Now we have two choices. First, we can assume that there is only one time-axis, which changes the 'present', but we cannot explain how the present changes. Second, we can postulate the existence of infinite time-axes, each of which changes the 'present' on another axis, and we still cannot explain how the present changes, because the cascade of such time axes is never ending.

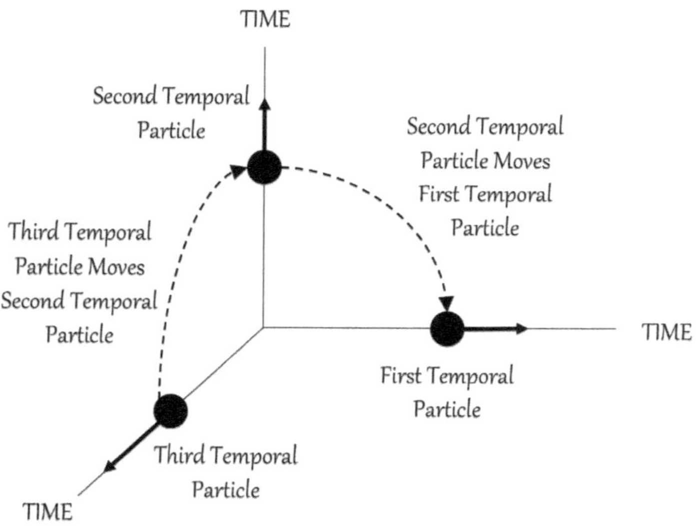

Figure-6 The Infinite Cascading of Temporal Particles

A solution that involves infinite assumptions and still fails to answer a question is worse than a solution that fails to answer the question with one assumption. The parsimony of assumptions dictates that we take the movement of time for granted. We can suppose that time moves automatically, and due to its passing, other things move. We can now focus on the explanation of the movement of material particles, but not on the explanation of time's passing. This works for most cases but not for the question of the origin of the universe. The question is: Does time pass if the universe doesn't exist? If time did not pass, then how could the universe come into existence from a state in which it was previously non-existent? If, however, time passes, then there must be something that exists even when the universe doesn't exist and causes its origin. To avoid the problem of infinite cascading time-objects, which are moved by the movement of other time-objects, this time would have to be *self-caused.* In short, it must be an object that doesn't itself move, and yet it causes everything else to move.

The Unmoved Mover

Aristotle called such an object an Unmoved Mover[3] and used it as a justification for the existence of God, as the cause of the universe. The idea of an Unmoved Mover supposes that God puts the world into motion once, after which the world is perpetually in motion. But this idea takes the issue of time for granted. When the existence of time is brought into focus, then the Unmoved Mover reappears in science as time. Now, we need to not only explain how an Unmoved Mover puts the world into motion at the beginning of the universe, but also how time itself moves perpetually. Since the idea of a time particle involves an infinite cascade of moving particles, therefore, the Unmoved Mover must be perpetually required. Indeed, since nothing would change if time wasn't changing, therefore, it would follow that nothing can change unless the Unmoved Mover wills a change.

But given that scientists are averse to God, or at least would not want God to meddle with science, the question of time's origin is mostly disregarded. Time is now treated as a *property* that can be measured like other physical properties (e.g. mass) using physical instruments. This is a problem because physical properties are *unique* to objects. For example, the mass of a particle is attached to a particle, and not to every other particle. If time is treated as a physical property, then it too must be attached to only one particle. And that would mean that every particle must have a unique sense of time, which contradicts the idea that time is common for all particles.

Treating time like a physical property is also not appropriate because there must be a cause of change in time, which then causes changes to other properties. If time is just like a position in space, and changes to positions in space require an explanation, then change to the position of a particle on the time-axis also deserves a similar kind of explanation. If we treat time like space, and only explain the changes in spatial positions but not the temporal position, then we have two ideas of explanation: in the spatial case, the position changes due to energy and force, but in the other case it doesn't. And yet, as we noted, given the aversion in science to self-cased agents, there is no other alternative but to disregard the true origin of time's passing.

The parameterization of time now leads to a definition of time as the movement of some clock. In short, we stop asking: What is time? How does it arise? What causes the passing of time? Etc. We shut up and measure.

The problem is that clocks are also moving *in* time rather than *moving time*. Time is as much a property of the clock—because the clock is an object—as it is of every other object that is not used to measure time. In fact, since every particle in the universe has a position in time, therefore, time is a property attached to every particle, and every particle must have an internal clock. Hence, no object can be used to measure the time for another object, just as no object's position can be used to decide another object's position. If the motion of the sun is treated as a clock, then it is a motion *in* time, rather than the motion *of* time. If the sun's motion is time, then that motion would be the cause of change to every particle's time. That would however mean that the entire universe is moving due to the sun's motion, and the causality between these motions would require another type of physical force between the sun and each particle. Without such a physical force, how the sun's motion causes time would be left unexplained.

In short, the parameterization of time brings many unsolved problems. First, we cannot say that time is a common property for all objects because each object is attached a property called time. Second, by this attachment, every object becomes a clock for its own time measurement, and we cannot use a standard clock even to measure time. Third, we are still left with the original problem of how each particle's time changes, because unlike the change in the spatial position, the change in time has no explanation.

The Problems of Energy

Given the problems associated with time's parameterization, we are compelled to reconsider the viability of the Unmoved Mover to explain the continuous flow of time. But this reconsideration in case of time is more problematic than the Unmoved Mover was in Aristotle's philosophy. In Aristotle's philosophy, the Unmoved Mover had to put the universe into motion by assigning every particle in the universe its momentum and position in space. This requires God to create infinite amounts of energy from nothing, and therefore it contradicts the principle of energy conservation[4]. But this is not the only problem when the Unmoved Mover causes time.

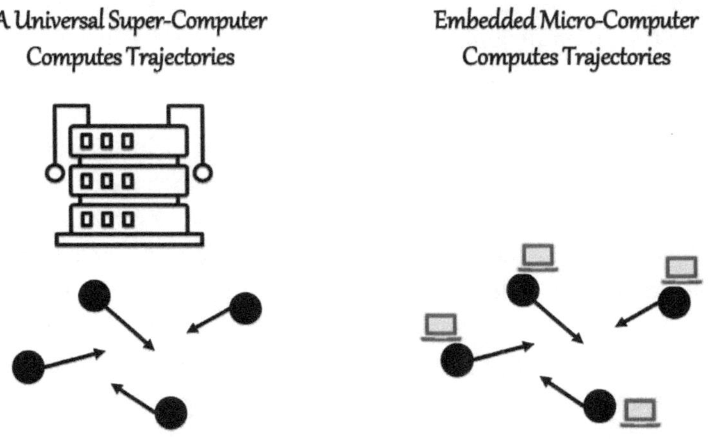

Figure-7 Alternative Computations of Natural Laws

Further problems arise when we consider questions about the computation of the mathematical laws of nature[5]. We can imagine two possibilities. First, that each particle has a sense of flowing time, which then drives the computation of the mathematical laws, which then determine the subsequent state of the particle. However, since computation of the mathematical laws requires some energy, therefore, these computers would eventually come to a halt, unless they were constantly being fed energy. As the computation halts, the particles will also come to a halt since they do not have the energy to compute the next state. Second, we can postulate the existence of a Universal Computer[6] that computes the evolution of each particle, but the result is similar—continuous computation requires energy, and in the absence of energy, the computation halts, and time comes to an end.

Thus, the Unmoved Mover for time brings two key problems. First, each particle in the universe must be put into motion by giving it some energy, and that requires an infinite amount of energy. Second, as these particles exert some forces upon each other, the computation of the effects of these forces requires energy, which also turns out to be infinite. If the first energy is supplied by God, but the second is not, then the universe would not be governed by laws; there would simply be energy but the next state would not be determined by the previous state according to some laws of nature. If instead the second energy is supplied by God, but not the first, then the universe would never change, because the input to all laws is always zero. Therefore, both types of energies must be supplied, and that entails an (1) infinite investment of energy at the beginning of the universe, and (2) a continuous investment of infinite amounts of energy to keep the universe moving according to the results of computation of some natural laws.

When time is parameterized, then change is attributed to physical properties, and time has an incidental role in change. For example, in the gravitational theory, the gravitational force—exerted by mass—is the cause of motion, and time is a parameter in equations of motion. Therefore, in a physical theory of nature, matter is the cause of change. But in the mathematical model, time is the cause of change: i.e., while performing computations, we must substitute the parametric time with successive values. But while trying to explain the trajectory, we must employ ideas of 'mass' and 'force'. The former leads to the problem of infinite energy in computation, whereas the latter leads to the problem of infinite energy in matter.

We must bear in mind that time isn't just *passing* but its passing is the cause of the computation of the next state of a system[7]. If time stops, then the computation halts, and the next state comes to a halt. If we say that time is merely a parameter, then the computer must also be working *in time*, and its working would require another computer to produce time, which again leads to an infinite cascade of time objects. In either case, the existence of time necessitates an Unmoved Mover, but now, with the necessity of law computation, the Unmoved Mover is also either an Unmoved Universal Computer, or infinite numbers of Unmoved Embedded Computers. In both cases, the Unmoved Mover is required to produce infinite amounts of energy. And this necessity makes the Unmoved Mover unacceptable.

The Self-Moved Mover

Since the idea of an Unmoved Mover proves to be so impractical and counterintuitive to all ideas of energy conservation in science, therefore, it is more prudent to think in terms of possibility and choice, rather than force and energy. We must still solve the twin problems of change we discussed above, but now these problems are much simpler: (1) the initial change is not adding an infinite amount of energy to set the universe into motion; the initial change is the choice of the next event, and (2) the subsequent changes don't require the computation of how energy is transformed from one state to another, because factually the energy is never transformed from one state to another because all states of energy are merely possibilities[8]. There is still a need to determine the next state, which is not a transformation of the previous state but a *choice* of a new state. The new state may not have the same energy, mass, charge, momentum, angular momentum, etc. as the previous states, because all these parameters simply describe a state, and all such states are eternally a possibility. Since they are not created or destroyed, therefore, the 'next state' is not about the *evolution* from the previous state, but merely the *selection* of the next state. Thus, when matter is defined as a possibility, the problem of computation disappears. Of course, since the new states may not have the same energy, therefore, we would say: some of the energy that was previously visible has now become *potential energy*.

Thus, we dissolve both problems of energy as they exist in the case of the Unmoved Mover, because we change our idea of causality from the computation of natural laws, or pushing of the world by force, to the evolution of the world through the combination of possibility and choice.

Given the remarkable differences between these two ideas (of the computation of natural laws vs. the selection of possibility), I will distinguish the Unmoved Mover (that sets the universe into motion and then constantly moves it through natural laws) from a *Self-Moved Mover* (that selects the future states from a domain of possibilities, at the start and then perpetually). In the case of an Unmoved Mover, there is a time-axis with a time-object on that axis, which is moved along the axis by an Unmoved Mover. In the case of a Self-Moved Mover, there is no time-axis, and no time-object that moves on the time-axis, but time is the choice of the next event made by the Self-Moved Mover. Indeed, the idea of a time-axis and a time-particle are totally superfluous due to the infinite cascade of such axes and particles For reasons of optimality, we must thus reject the existence of a time-axis and replace it with the Self-Moved Mover who moves the world by a choice. This choice is not made once but made perpetually to create time. Energy is not expended in this process because such expense was needed to compute laws, and laws are no longer being computed. Hence, the *Self-Moved Mover* doesn't suffer from the same issues as the Unmoved Mover.

Responsibility as a Natural Principle

But we are now faced with a new question: If the next state of the universe is a choice, does it mean that there are no laws that govern the next state? Alternately, do the choices of a Self-Moved Mover determine the next state of the universe randomly? Doesn't randomness follow all choice?

Choices are never random if we are driven toward goals, and choices can be made only if we are driven toward goals[9]. Therefore, choices cannot exist without goals. To say that we are free to choose is identical to saying that we have a purpose. What is purpose? It is the quest for happiness. All purposes are simply different methods of attaining happiness. However, to be happy, we must have the ability to choose what we desire, and this creates a tradeoff because if choices have effects and consequences that modify the possibilities, then bad choices would make long-lived happiness impossible. Thus arises the idea that the purpose of existence is happiness, and we are free to choose what makes us happy. However, the choices for happiness must also be responsible since irresponsible choices reduce happiness. Thus, in one sense, choices can be arbitrary since we choose our happiness. But, in another sense, choices cannot be arbitrary because there are *effects* and *consequences* of choice. Choices become random if there are no consequences. And randomness arises due to the ignorance about the effects and consequences of choices. If we are aware of the effects and consequences of choices, the choices would not be random; they would rather be goal-driven toward the maximization of happiness, but also not reckless.

We can apply these ideas to the choices of the next state of the universe. The application will say that the universe exists for pleasure. But maximization of pleasure also means that choices must be responsible. As a result, all kinds of pleasures and pains must exist in the universe—the pleasures because the universe exists for enjoyment, and the pains because it is essential to correct the irresponsible choices, for maximizing the long-term pleasure! Obviously, this means that the pleasure of the universe is not unlimited, and choices must be responsible precisely because the universe is finite. For instance, if the universe were infinite, then there would be no limit to pleasures; we could go on consuming whatever we want, and the supply would never diminish. Then, we would never have to trade-off short-term happiness for long-term pleasure. The finitude of the universe means that resources must be distributed equitably, and consumption must follow regeneration, which can follow consumption, since the universe is finite. Hence, we can tie together three distinct ideas that create a coherent worldview—(1) the universe is finite, (2) the universe exists for pleasure, (3) the finitude of the universe entails that to get pleasure, the choices must be responsible.

The ideas of finitude, choice, purpose, and pleasure are fairly straightforward, and need no explanation. But the idea of responsibility is not so obvious. What is responsibility when we are free to pursue pleasure? The simple answer is that pleasure must be restrained; even if we are going to enjoy, there must be a method to the madness. The restriction on pleasure arises because the process or path to happiness must be governed by some principles. We will discuss these principles in the next chapter, but briefly, every value involves some cost, every output requires some effort. The basic principle is that the costs and efforts should be minimized while the value and output are maximized; however, the minimum cannot be zero.

Nature, even in modern science, is governed by *least action* principles that lead to shortest path trajectories. In effect, the goals are not achieved without effort, but they also don't require unreasonably excessive effort. Thus, the relation between choices and possibilities is established by a bi-directional *responsibility*. The possibility exerts a responsibility on choice—*if* choice is misused, then there are natural consequences, and the proper use of choices is incumbent upon the person who chooses. Likewise, choice exerts a responsibility on possibility—*if* choice is at all to be used, then the results, effects and consequences must be decided *fairly*. What is fairness? It is the simple idea that loss and gain are equally balanced. To gain something, we must lose something, but what we are losing is not lesser or greater than what we are gaining. As a result, the universe cannot be only gain, or only pleasure. It must involve some sacrifice or loss to obtain some pleasure.

In short, possibility exerts a responsibility on choice—the choices have results, effects, and consequences. Similarly, choice exerts a responsibility on possibility—the results, effects and consequences are determined fairly or equitably. There can be no choice without responsibility, and there cannot be any responsibility without choice. Hence, the principles of choice and possibility must be connected through a bidirectional responsibility.

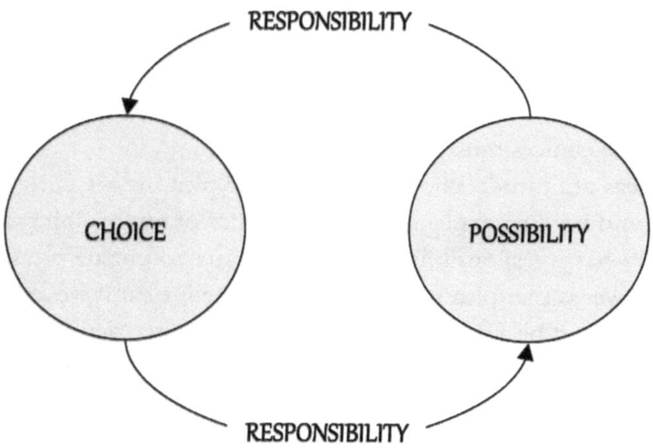

Figure-8 The Choice-Possibility-Responsibility Relation

This connection between choice, possibility, and responsibility is pervasive in our world. For example, if you become a citizen of a country, then you must bear the responsibility of following the country's laws. But that doesn't mean that the government can create arbitrary and senseless laws. Indeed, you will choose to become a citizen only if you know that the laws of the country are fair. Thus, most people want to become citizens of those countries where the laws are fair. By this measure, most people try to escape autocratic, irrational, and unfair regimes. Thus, the citizens choose a country based on the fairness of its laws, and the country chooses citizens based on their readiness to follow their laws. Thus, citizenship asserts bidirectional responsibility, and the citizens are therefore responsible for following the laws, and the government is responsible for making the laws fair. If laws are frivolous, then choice disengages from the possibility. Likewise, if choice is frivolous, then the possibility reprimands and punishes the choice.

It might seem that we are going too far from the questions of time, but we are not. I am trying to describe a paradigm that substitutes the problematic ideas of matter governed by mathematical laws with an alternative, free of these problems. The alternative involves choice, possibility, and responsibility. Choices create randomness when there is no responsibility. Likewise, possibility creates unwanted complexity without responsibility. When bidirectional responsibility is asserted, then nature has order in *how* things are done, without undermining the goals of *what* nature's purpose is. Choice denotes the existence of a purpose in nature, and order is the regularity in the natural processes through which the purpose in nature is achieved. This order is free from the opposing problems of determinism and causal incompleteness; it is also free of the problems of energy creation, infinite cascade, etc. Thus, we can see its merits in two ways: (1) it is free of the *scientific problems* of causation, and (2) it is free of the *human problems* of choice.

Two Interrelated Realities

With this idea of choice, possibility, and responsibility, we can now speak about the twin realities of space and time in a new way. Space is the domain of possibility, which exerts a set of simple rules of responsibility upon choice. Time is the domain of choices, which engages with this possibility because the rules of engagement are defined as reasonableness. The world is produced by the combination of possibility and choice, and not by the exertion of force or the creation and the annihilation of energy.

The domain of possibilities—because it exerts a choice—can be called the *Self-Mover Moved*[10], as it is moved by a *Self-Moved Mover*. The domains of possibility and choice are therefore self-driven, but one is a Mover and the other is Moved. Since the Self-Mover Moved is moved by a Self-Moved Mover, hence, causality must be attributed primarily to the Mover rather than the Moved. But that doesn't mean the Moved is inert. It means that choice has the prime prerogative to engage with a responsible possibility. The possibility controls the choice by creating results, effects, and consequences. And the choice controls the possibility by creating changes by its decisions. Thus, we cannot separate choice and possibility, if the relation between them is a responsible use of choice, and a responsible set of rules that deliver results, effects, and consequences. They are thus *inextricable*.

For simplicity, we can call these two realities *God* and *Nature*. God is the predominant entity of choice, and acts as time to select the future state of the universe. Conversely, Nature is the predominant entity of possibility. But Nature is not unconscious, inert, impersonalized, or devoid of choice. Nature is rather the choice of the principles of reasonableness that lead to what we call in science "natural laws". The laws could be different, but they are what they are; so, there is a choice of laws, and that choice of some laws over the other laws is Nature's choice. But this choice of laws is not random; it is the reasonable precondition for choice to engage with possibility.

In the Unmoved Mover case, God creates Nature, but in the case of the Self-Moved Mover and Self-Mover Moved, both God and Nature are eternal. They are also separate because the Mover may not engage with the Moved, and the Moved may not entertain the Mover. Their engagement is their responsible choice. God exists as Time, and this Time passes as the succession of God's choices, which select a subset of possibilities, and the possibility evolves as the effect and consequence of God's choices. This choice, which controls nature, defines the goal of nature. Nature on the other hand determines the processes by which a goal is attained. The goals are reasonable and the processes of attaining the goals are reasonable. Thus, the world can be known rationally, but rationality is not the uber principle. The uber principle is the precondition for choice and possibility to engage through a choice of reasonable goals, and a choice of reasonable processes. Thus, rationality doesn't contradict choice; rationality refers to choices.

Most scientists consider the questions of "Why does the universe exists?" "How is it governed by some precise laws?" and "How are these mathematical laws governing the evolution of a physical world?" meta-scientific questions: they are outside science. Since these questions are rejected from science, they become part of philosophy. But that doesn't make the questions less important since the questions are relegated to philosophy only because science is unable to provide satisfactory answers to them. Therefore, we must disabuse ourselves of the idea that these are not important. They are not important to science, *because* science cannot answer them. This is not the assertion of a rational principle; it is the assertion of choice.

The Role of the Observer

If we add the observer to the realities of God and Nature, then apart from *when* and *where* we also obtain the notion of *who*. The interaction of God and Nature produces a world, but the observer has the choice to determine which subset of the world the observer passes through. The choices are also subject to the conditions of responsibility. In short, the observer is not compelled to be part of the world-experience, and that simply means that the observer is different from the world. However, if the observer enters the world, then he must also act responsibly. If irresponsible choices are made, then the possibility that he can choose from changes (to undesirable alternatives) and contracts (i.e. eliminates desirable alternatives). If, on the other hand, responsible choices are made, then then possibility he can choose from changes (to desirable alternatives) and expands (i.e. allows undesirable alternatives). Nature thus *tempts* the observer to act irresponsibly and then *punishes* him for irresponsible action. If the observer is punished, then he corrects his choices, and his possibilities change to desirable ones, but they also expand into permitting irresponsible choices. If the observer misuses the choices for enjoyment—as the choices have expanded into irresponsible alternatives—then he leads himself into another round of undesirable situations, correction of choices, and then pleasure. Thus, the Nature-God combination creates a *cycle* of good to bad and bad to good.

In subsequent chapters, I will discuss how time is itself cyclic in the sense that the world—independent of all observers—evolves in a repetitive manner. But before we get there, we can also understand how each individual observer exists in a cycle of change, resulting from his choices.

Choice, possibility, and responsibility are therefore central constructs to solve the problems of passing time. Since choices exist in the Mover and the Moved, besides the observer, therefore, each of them is a consciousness. However, they are also different in how their choices are effected. God's choice amounts to *when* and *what*—because God determines the purpose in nature, and thereby decides what happens when. Nature's choice amounts to *where* and *how*—because the same goal can be achieved in many ways, and Nature chooses the most optimal, parsimonious, simple, and effective solution. An observer's choice amounts to *why* and *who*—do I want to participate in the world, and why am I subject to experiences that I don't desire?

Since the questions of what, how, and who[11] are irreducible, therefore, the answers are also distinct. Thus, any attempt to reduce these three realities to a fewer number of realities must fail in answering some questions, which would in turn constitute the *incompleteness* of a scientific theory. Conversely, since these are the only possible questions, therefore, the answer to all these questions constitutes the completeness of a scientific theory.

Now, we can evaluate the solution to the problems of time experience based on rationality—the solution is *consistent* and *complete*. The consistency is given by the existence of three ontologies, so even though they are all consciousness, the differing answers to the nature of consciousness do not create a contradiction, because they are different *types* of consciousness. The completeness is given by the fact that the six types of questions are all that can be asked, and therefore, their answers address all the questions.

Takeaways from Time's Mechanism

The discussion about how time passes and the solution to the problems it poses, therefore, leads us to some key ideas that will prove useful throughout the book. We can quickly summarize these ideas briefly here:

- We must understand that time's passing presents problems of continuous and infinite energy production to keep the universe in motion. These two problems make a scientific contention about energy conservation untenable, which makes all other laws untenable.
- There are unanswered meta-scientific questions such as "Why

does the world exist", "Why is it governed by certain fixed laws?" and "Why are these laws subject to simplicity, parsimony, and optimality?" Without an explanation, science itself is incomplete.

- The postulate that nature is created by the combination of choice and possibility—which we uncovered in the previous chapter through the ideas that we can have goals and memories—can also be used to answer the other difficult questions posed above.

- The solution to these problems, however, requires us to add the idea of *responsibility* to the previous ideas of choice and possibility that we discussed in the previous chapter. Responsibility, however, goes both ways—choices must be responsible, but the laws that determine the effects and consequences must also be responsible.

- The lawfulness of nature can now be explained as the consequence of the principles of reasonableness in nature—i.e. every gain involves an equal loss, and vice versa. This reasonableness can also be called the fairness involved in the fulfilment of one's purpose.

- Like we discussed in the previous chapter, we are again led to three key ideas of time, space, and observer. Additionally, time is now identified as God, and space is understood as Nature. Since each of them is capable of choice, therefore they are all consciousness.

- The contradictions between matter and mind, or free will and natural laws, can be rejected. The solution stands the tests of rationality (consistency and completeness), scientific knowledge (causal explanations) and human needs (choices and responsibilities).

3

Do the Past and the Future Change the Present?

The Least Action Principle

The laws of classical mechanics can be formulated using *least action principles*: particles move along the *shortest path* to a destination[1]. The trouble is that before the shortest path is decided, a destination must be known, which implies that the future must exist conceptually at the present. The necessity of the future's existence makes the least action principle contradictory to the deterministic formulation, because in the latter case, the future doesn't exist presently. Nevertheless, this conceptual difference between the deterministic and the least action formulation of classical laws remains obscured due to three reasons. First, we use the principle to decide the destination based on the shortest path (when it was intended to decide the shortest path to a destination). Second, there are many destinations given by a shortest path since the space of classical physics is uniform and moving from one point to another is not harder in one place and easier in another[2]; therefore, the uncertainty in the destination based on the shortest-path is resolved by supposing that a vector force pushes the particle in a certain direction. The vector force is additionally required aside from the least action principle. Third, even if we postulate the presence of a force, which constitutes *dynamics*, the *kinematics* remains uncertain because during an interaction, particles can split and join, and energy can be redistributed in many ways; this is when the equivalence of least action and deterministic formulations of classical mechanics fails. But the situation can still be salvaged if we say that the *system* follows the least action principle, even if the parts of the system are not deterministic. However, to apply the least action principle, we must know the destination. Thus, the difference between the least action and deterministic formulations of classical mechanics is visible only when we consider particle collections instead of individual particles, and the failure of classical determinism can be corrected by the use of the least action principle.

To understand this problem, consider the different possible outcomes of the collision of a moving particle with a stationary particle. First, the entire energy of the moving particle can be transferred to the stationary particle; if the stationary particle is heavier, then it will move slower compared to the moving particle, and if the stationary particle is lighter, then it will move faster compared to the moving particle. Second, the energy of the two particles can be distributed between the particles in many ways; for example, both particles can move at different speeds and in different directions (subject to the total energy and momentum being conserved). Third, the collision may coalesce or split the particles, with each particle having a different energy afterwards. The laws of classical physics do not predict which one of these innumerable possible outcomes would be observed[3]. Indeed, since each of these numerous scenarios are equally permitted by the laws of physics, therefore, each of these can be the prior scenario with one of the others being the subsequent scenarios. As a result, we also cannot explain the outcome, even after the fact, because there are innumerable explanations[4]. When the predictions and explanations fail, then a theory becomes *incomplete*, since the goal of science is to predict and explain observations.

This is when we begin to clearly see the value of the least action principle: it can make the classical predictive framework complete because it involves two principles—a destination and a trajectory toward the destination—rather than one principle of using the current state to determine the future state. The future state is now the destination, and the shortest path principle determines the trajectory toward the destination. Since the postulate of a destination requires the assumption that the future exists now, and it can be chosen, therefore, the predictive and explanatory failures of classical determinism would entail the need for an alternative viewpoint.

This problem can be generalized as follows: The predictive and explanatory laws of classical physics permit innumerable distributions of matter and energy, and thus such laws are incomplete[5]. To pick one out of these infinite possibilities, we need a choice. That choice, however, doesn't entail a collapse of classical laws any more than they have already failed. This is because the choice of a destination decides the future state, whereas the laws of classical physics determine the shortest path to achieve that state.

Destinations and Trajectories

The new problem is that there are two kinds of evolutions and hence two kinds of time—pertaining to the evolution of destinations and of trajectories. There is a time in which *what* will happen evolves, and there is a time in which *how* it will happen evolves. These two temporal evolutions pertain to the time we observe, as they construct the different trajectories. As the future evolves, it naturally influences the trajectories used to reach the destinations, and therefore, the future must determine the present.

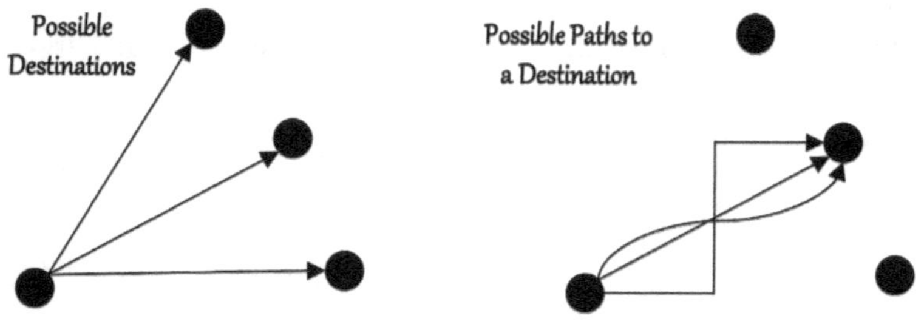

Figure-9 The Choice of Destinations and Trajectories

However, if we alter causality to allow for the future being selected by a choice, then it seems that any future can be attained from any present, and that idea must now contradict the notion that the present determines the future. We are again left with unidirectional causality. If determinism fails to predict one of the many possible outcomes, then the system of choice predicts even those outcomes that might never occur. In short, we have traded off the incompleteness of the former system with the inconsistency of the latter system, and both solutions seem to be equally problematic.

To address these issues, we must change the idea of space. Space presents possible destinations (so the future exists at the present), but all possibilities are not equally accessible. Some possibilities are attained with lesser effort, while others require greater effort. In classical physics, such barriers are represented by *force fields*, which also act on objects trying to reach a destination. A force field, however, affects everything in the universe. An electromagnetic field, for example, will attract or repel charged particles even if they are not trying to reach a destination. A *possibility barrier* is instead like a wall surrounding a house, which prevents access to a house, but has no effect on those outside the wall. Therefore, a possibility barrier seems like a force field, but there are obvious differences between them. If space is redefined as simpler or harder possibilities[6], then we recover the idea that the future is also determined by the present, because destinations are chosen based on whether they are accessible, in contrast to the choice of a path suited to reach a destination. The former decides the future based on the present, while the latter determines the present based on the future.

Change Through Optimization

When the space of possibilities is not uniform, then more than one trajectory to a destination is possible, and these trajectories can involve different notions of optimality. For example, if you were trying to reach the top of a cliff, you might consider two paths—one that involves a longer drive and a shorter climb, whereas the other involves a shorter drive and a longer climb. If you were short on gasoline, but abundant on time, then you would prefer the alternative that conserves gasoline at the expense of time. But if you had abundant gasoline, and less time, then you would prefer the trajectory that involves a longer drive followed by a shorter climb. In effect, the principle of optimality is universal, but its application varies with the situation, because we assign different *priorities* to different parameters in different situations. As a result, when we are presented with more than one alternative trajectory to a destination, one of these paths is always chosen based on the principle of least action; however, the principle of least action requires determining what we consider the 'action' in each situation.

We understand that every choice involves a compromise: something must be traded off to attain something else[7]. We understand this idea incorrectly as the principle of energy conservation in which energy has many *forms*, and some forms are more valuable than others. The problem is that one such form of energy is that it exists as a potential and remains unobservable. How would we know that the energy is indeed present, but unavailable? Therefore, even if we say that the total energy is conserved, in practice, the total *usable* energy is not conserved (this problem is very precisely illustrated in thermodynamic phenomena where a property called 'availability' is associated with energy; we will discuss this more fully in a later chapter). Since the different forms of energy can have greater or lesser usability at different times (if they have become potentials), or greater or lesser usefulness (for our goals), hence, all forms of energy aren't equal.

For example, if you wanted to have more gasoline after a mountain climb, you might decide to run, instead of walk, on the longer path, to preserve the gasoline as well as save time. Likewise, we might possess some forms of energy more than the others, based on the past trajectory.

Therefore, the past determines what kinds of energy is usable, and the future determines what kind of energy is more useful. If greater and lesser priorities to usefulness, and greater and lesser quantities to usability are assigned to different forms of energies, then the optimization principle can be applied to the usability and usefulness: something more useful but with a lesser usability must be conserved the most, and the something more usable but with lesser usability must be conserved the least. Now, we are not trying to minimize the *energy* consumption but trying to optimize the *value* attached to this energy. Optimization means that the costs must be minimized while value must be maximized. A physical transaction now begins to mimic the properties of an economic activity because: (1) we trade off some value to get another value, and (2) we try to spend less value and obtain more value, but the terms 'more' and 'less' are relative to a situation and a goal. The least action principle now represents the *minimization* of costs. It is not sufficient, and it must be complemented by the principle of *maximizing* the value.

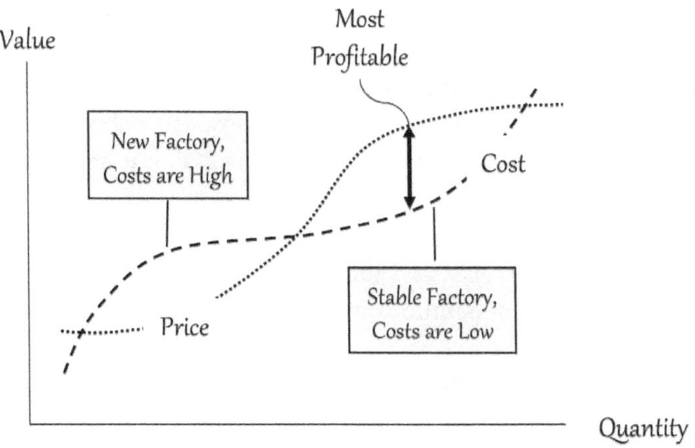

Figure-10 The Changing Cost-Price Curves

Figure-10 illustrates this idea through the simple example of a factory that produces some widgets. When a factory is setup but hasn't attained its optimal quality or quantity of production, then the costs are high, and the prices are low. As the factory matures, the investments in its operations decline, but the production and the quality ramp up and costs are now reduced but the prices are increased. The factory owner seeks an operational point where the profits are maximized, *and* the costs are minimized. No owner tries to individually maximize profits or minimize costs. Therefore, we can note two key necessities: (1) we must consider both value maximization and cost minimization, and (2) the goals is decided by that point of optimality where the cost-value difference is maximized, if one can afford the cost. What we can afford comes from the past—i.e. the energy that was saved (for the moment we can suppose that we are not borrowing energy from others). And the value we can obtain comes from the future—i.e. the energy that we can generate by expending the saved energy from the past.

The least action principle is not wrong, but it is only *half* the story—the minimization of costs. It assumes the existence of a destination at the present, which constitutes the other half of a story. If the destination is not known, then cost minimization is not sufficient. Completeness is obtained only if the destinations are determined by the other half—i.e. the maximization of value. Both costs and values can be objective but contextual—i.e. the cost and value depend on a context, but common for all observers. Similarly, both cost and value can be subjective but follow value thresholds (i.e. more valuable below a threshold, and less valuable above a threshold). Therefore, the principle of optimization is universal, but its application

(1) Is subjective and contextual,
(2) Depends on the past and the future, and
(3) Requires both maximization and minimization.

The least action principle works on the premise that

(1) Only costs are considered,
(2) Only minimization is required, and
(3) The costs of reaching a destination are objective

The net result of using least action principles is that the theory remains incomplete, the role of the future in deciding the outcome is disregarded, and the potentials acquired in the past—e.g. as skills to create change—are disregarded because only the present is considered relevant to the future. These are simply the results of viewing the world physically and deterministically. If instead we think of the world as possibility and choice, then, we can appreciate why the process must be more nuanced and complicated.

The Use of Variational Principles

The existence of choices, therefore, doesn't undermine the importance of *rationality*. However, they change its role. The old role of rationality was to determine the future from the present, which is unachievable because the past is consistent with innumerable futures, and vice versa. The new role is that rationality serves to determine the most optimal path to a chosen future, and the most optimal future based on the present conditions. Each such path has a 'cost', and reason minimizes the costs of the path, and this cost minimization is the least action principle. However, the minimization must also be complemented by the maximization of value. The goals are selected based on maximum value, and trajectories are decided based on minimum cost. Values, however, are subjective while costs are objective. But whether we want to attain that value cannot be determined by the lowest cost. Therefore, we must employ a general class of *Variational Principles*[8] that perform not just cost minimization, but also value maximization.

Now, we can combine the variational principles with the ideas of space and time as possibility and choice discussed in the previous chapters. The result of this combination is that space makes certain goals easier or harder to attain, and that is a consequence of the structure of space. This structure appears as the *curvature* in space which can be compared to a land with mountains and valleys. In short, the space is not *flat*. Remember, however, we are speaking about the space of possibilities, and not physical space employed in classical physical theories. The curvature of space simply means that each possibility requires a different cost, and the least cost principle can be used to find the shortest path to a possibility. Whether that possibility is desired depends on the choices of goals. And whether cost can be borne depends on the available energy acquired through previous states. Not everyone desires all possibilities, and not everyone has the energy in the right form that can be used to attain the desired possibilities. Therefore, everyone must employ a Variational Principle of optimization which maximizes their value while minimizing their costs—*if* they are behaving rationally.

We must recall that rationality is optional when choices are permitted. Rationality is the awareness of results, effects, and consequences of choices, but not everyone knows of these outcomes, and hence they can act irrationally. Irrationality is thus not forbidden but it produces adverse outcomes, and therefore irrationality can be *proven* to be irrational, but it can also be *chosen*. Thus, reason may not determine our choices, although reason determines the *best choice*. Innate to choices is the idea that we may not make the best choice. Thus, the observers are free to be rational or irrational—i.e. they may not choose to maximize the value and not minimize the costs.

Conflict in Variational Principles

If this was not complex enough, then there are further nuances to the use of variational principles. One such problem is that unlike the least action, which minimizes a single entity called 'action', in complex scenarios, there are many ways in which the most optimal path can be determined. For example, the reduction of 'costs' can be achieved using two principles—simplicity and parsimony—but these are not necessarily consistent.

To illustrate, most people who file tax returns at the end of the year find the tax forms overly complicated because most of the questions in a tax form do not apply to most people, and yet we are required to answer them. This problem could be solved by creating many types of tax forms, each with fewer (although different) questions suited for different kinds of taxpayers. Now, we get a question: Are we shifting the burden from trying to fill a more complex form to figuring out which form must be filled? If fewer tax forms are used, then the forms are parsimonious, but each form is not simple. Conversely, if we simplify the forms, then we also have more forms, and the forms are not parsimonious. Should we have fewer complex forms or more simple forms? Notably, both options will reduce costs, but they will also shift the costs from one type to another. If a nation had a hundred people, then asking them to fill a single more complex form would be cheaper than employing 5 people to design simpler forms. But if a nation had a million people, then employing 5 people would be a lesser cost. This means that optimality needs adapting the principles of simplicity and parsimony; they seem to be the same idea, but they can be contradictory.

That doesn't disprove the principles, however. For example, even if the government were trying to create many simpler forms, it would not try to multiply these forms beyond necessity. Therefore, the principle of parsimony still holds good, and yet, it may be prioritized lower over simplicity. Conversely, if the government prioritized parsimony over simplicity, then it could create a fewer tax forms, but not overcomplicate them beyond necessity. In short, the principle of simplicity would still hold true, although it would be prioritized lower than the principle of parsimony. The conclusion is that we must uphold both the principles of parsimony and simplicity, but they don't always have equal priorities. The universality of the principles doesn't contradict the contextual applications of such principles. This gives us a better understanding of the nature of choice: it is the selection of the more important principle, from a collection of universal principles.

Three Kinds of Rationalities

We must now try to connect the contextuality and individuality of choices to the principle of responsibility of choices that we discussed in the last chapter. The principle of individuality or choice is that (1) we can choose to be irrational, and (2) we can choose to act irresponsibility. The principle of contextuality is that in each situation we *must* make the right priority calls about what is less or more important. For example, sometimes we must prioritize simplicity over parsimony, and sometimes we must do the reverse. Effectively, contextuality defines *responsibility* and when individual choice violates the principle of contextual prioritization, it also constitutes *irresponsibility*. In simple terms, responsibility means that we have a contextual duty, and individuality means that we have the freedom to follow the duty. Once a principle is prioritized, then according to our rationality, a *result* is produced. But according to the responsibility, if we follow the duty, then Nature's effects and consequences will expand our freedom, and if we don't follow the duty, then Nature's effects will contract our freedom. Thus, our rationality pertains to the *results* while Nature's rationality delivers *effects* and *consequences*. Our rationality is important in using the context for desired outcomes, and we are free to desire—even if our desires are irresponsible. However, Nature's rationality and responsibility must always be applied, and it determines the future effects and the consequences of our choices.

There are hence three kinds of rationalities. First, the *universal* principle of rationality is ideas such as fairness and optimality; fairness means costs and values are equal; optimality means simplicity and parsimony. However, these are principles, rather than laws, in the sense that they can be violated. Second, the *individual* principle of rationality is that we are free to choose our pleasures, determine what we like or dislike. Third, the contextual rationality is that in each situation the principles of fairness and optimality must be applied differently, by assigning them different priorities. If these priorities are misapplied, or applied as if they were universal, then choices lead to irresponsible action, and the original contract of responsibility with Nature—i.e. that choices must be responsible—is broken. Now, we must bear the adverse effects and consequences of the irresponsible choices.

If a rational path is chosen, and responsible choices are made, then favorable effects and consequences would be obtained. But if an irrational path is chosen, and irresponsible choices are made, then adverse effects and consequences must be obtained. This means that Nature is always rational and responsible, even if *we* are not. If our rationality is identical to Nature's rationality, then the world would seem to behave according to our rationality. However, if our rationality deviates from Nature's rationality, then Nature produces effects and consequences against our irresponsible choices. When Nature doesn't behave as we had anticipated, then we are required to change our rationality, and have it conformed to Nature's rationality. That means respecting the original contract of choice with responsibility.

Time and Free Will

The least action principle leads to a denial of choice when (1) no future state is more desirable, (2) the space in which change occurs is uniform and isotropic, and (3) the past doesn't condition us into the abundance or shortages of different forms of energy, because energy has a single form. These are the assumptions of classical physics, but they are not the reality.

The rejection of these assumptions leads to the idea that there are two kinds of times—of changing goals and of trajectories—in which the past and the future decide the present, as goals are chosen based on the minimum costs, and costs are minimized based on goals. This is a far more complex process of optimization of different principles than we are accustomed to in classical determinism. The predictive and explanatory system is complete, but it uses universal principles, contextual differences, and subjective choices. In short, *universality* is now also complemented by *individuality* and *contextuality*. The universality is that universal principles such as simplicity and parsimony are applied by everyone. The contextuality is that in different situations, these principles must be applied differently by almost all people (assuming of course that they can apply these principles). And the individuality is that some people may still act irrationally in some situations.

The conclusion is that the classical caricature of time as a parameter that moves from the past to the future is incompatible with the lived experience of time. Unless our goals are illusory, or we have no rationality to optimize and simplify, or energy doesn't exist in many forms, the classical caricature of space and time is also false. Arguments against free will do not see how classical mechanics is indeterministic about the future outcomes, and incapable of explaining the emergence of the future from a fixed past. Hence, we can appeal to the failure of determinism in science, and we must appeal to the conflict with the lived experience to reconceive time.

But when we try to address the issues arising out of the failure of determinism in classical physics, we see complex problems of causation, because determinism is replaced by choice and possibility; their interaction requires responsibility; responsibility can be described rationally using universal principles, but these principles have to be contextually prioritized in a specific manner to constitute responsible choice, although they can be prioritized irresponsibly by individuals; that irresponsible prioritization leads to adverse effects and consequences; and that adversity then compels us to modify our rationality to fit Nature's rationality; when our rationality is thus adjusted, a temptation to be irrational is again created by nature, and the cycle of responsible action followed by irresponsible action repeats.

Takeaways from Effects of the Past and the Future

The discussion about the problems posed by the influence that the past and the future exert on the present, and the solution to these problems, therefore, leads us to some key ideas that will prove useful later in the book:

- Deterministic theories in classical physics present an incomplete model of change because the laws of nature are consistent with innumerable energy distributions. This incompleteness leads to the collapse of both *necessity* and *sufficiency* in the causal model.

- To solve this problem, we can consider the least action principle and how it was meant to determine the shortest path to a chosen destination, but to use this principle, we must be able to explain how the destination is chosen prior to deciding a trajectory.

- This problem requires us to widen the optimization problem into both minimization of costs, and maximization of value, and the least action principle pertains to the former, and must be supplanted by another principle for maximization of value. Together these two principles constitute a broader *variational principle*.

- Generally, variational principles assume that there is one quantity called an 'action' that must be optimized, but in reality, there are many such principles—such as simplicity and parsimony—which are often mutually conflicting, and contextuality and individuality determine which of these principles must be given priority.

- The use of a variational principle, subject to the choice of goals in the future and the potential abilities from the past, and the priority assigned to the 'actions' being contextually optimized, constitutes 'rationality'. It delivers the best outcomes, but it is not *necessary*.

- If choices incorrectly prioritize universal principles in some contexts, then nature will produce adverse effects and consequences; if correct priorities are employed, then nature will produce favorable effects and consequences. The definitions of favorable and unfavorable are based on individual likes and dislikes, but the receipt of such outcomes is based on the contract of responsible choice.

4
Does Time Pass Uniformly?

Each new little day slips out of my hand,
And then with another new day I stand.
But soon that is gone and folded away—
I wish I might keep forever one day!
I wish that one good day might always stay,
For the good days hurry on so fast,
Only the bad days seem to last.
—Annette Wynne

The Motion of Clocks

When time is reduced to the measurement of a parameter, a new problem arises: Is the clock moving uniformly over time? Or does it sometimes move faster or slower? If the clock moved faster or slower, then more or less time would *seem* to elapse. Should we consider a faulty clock that moves faster or slower an inaccurate representation of time? Or should we say that the clock is correct, and the rest of the world is moving slower or faster? Since we have no way of knowing which interpretation is correct, we also have no way of knowing whether some clock is indeed faulty. It follows that we have no way of knowing whether time passes uniformly or not.

If clocks sometimes moved faster or slower, then they would also gain and lose energy. The faster-moving clock, for example, would have more kinetic energy, and the slower-moving clock would lose some kinetic energy. In short, the non-uniform movement of the clocks would create and destroy energy. However, the problem is that even if the clock moves faster or slower, *we* will not know that it is moving faster or slower, because the measure of speed is given by the clock itself. Hence, the problem of whether time passes uniformly is in principle unsolvable *if* time is defined as the motion of a clock. This unsolvable problem is, in fact, converted into a 'law' called the *conservation of energy*. This law is conceptually identical to the notion that the time-axis is *homogeneous* or that time moves uniformly. Thus, the claims that time is homogeneous, and energy is conserved, are identical. But, truly speaking, the conservation of energy and the homogeneity of time, are just *hypotheses*, which mutually reinforce, and mutually imply each other. The energy conservation principle is a tautology because conservation is defined by the elapsed time, and time is defined by the conservation of energy. It's worth noting that I'm not insisting that the conservation principle is false. I'm saying that that it is not a 'law' that can be independently verified. It is rather an assumption upon which the rest of science is constructed.

Notably, energy conservation is not alone in the boat of tautological axioms. Other conservation laws such as that of momentum and angular momentum are also similarly tautological. The former depends on the idea that space is *homogeneous* and the latter on the hypothesis that space is *isotropic*. For example, if the distance between two adjacent points in space varied, then momentum would not be conserved because 'uniform motion' would entail that sometimes a particle moves faster, and sometimes slower. Likewise, if angles between two adjacent lines were not equal, then angular momentum would not be conserved, because 'uniform rotation' would mean that sometimes an object rotates faster, and sometimes slower. Thus the 'laws' of conservation of energy, momentum, and angular momentum in classical physics are not truly 'laws' but *assumptions* about the nature of space and time. There are no *a priori* reasons why space and time must be this way. However, since we cannot step 'outside' this space and time in classical physics[1], therefore, we also cannot independently verify the laws.

Now, you might say: Even if we cannot measure the uniformity of time using a clock, maybe there is another property which would vary based on a non-uniform time, and we could measure that property to decide if time is uniform. The short answer to that doubt is that space and time are assumed to be uniform and isotropic in all measurements, and upon this uniformity rests the measurement of every other quantity. For example, mass is measured by the movement of a pointer on a weighing scale, and to consistently measure mass, we must say that the distance between markings on the measuring scale is equidistant. Similarly, temperature is measured by mercury moving in a thermometer, which relies on space being uniform.

All physical properties are ultimately measured by the values of one of three properties—length, angle, or time. Mass or temperature measurements are not truly the measurements of mass or temperature; they are only the measurement of length or angle. Likewise, speed measurement is not truly a measure of speed, but the measurement of distance and time.

Problems in Deciding Uniformity

This problem can be solved if we say that the observer is outside space and time, and can, in principle, measure the uniformity of space and time. To make such claims, we would have to say that the observer is not material. But very quickly, additional problems arise. First, how does something outside space interact with something inside space? This would be the counterpart of the mind-body problem in Western philosophy. Second, what if every observer reported different values of distances and durations? After all, if the space and time that we presently use in science are not homogenous, then any other space and time used by observers could also have the same issue, leading to an infinite cascade of observers, each measuring the other's instruments and the conclusion would still be: We don't know!

Since time is measured by clocks, an additional problem arises in the measurement of time if each clock moves at a different rate *in different places*. It is not enough to say that each clock moves at a constant rate, because there are potentially infinite such clocks. Clock consistency is also necessary if we are going to repeat the experiments and apply the same laws of nature in different places (a "law" of nature would be useless if it were true only in one place). If clocks at different places move at different rates, then you can send a light beam through space—e.g. between two mirrors—and measure the time it takes to return to its source. If time moves non-uniformly, then depending on where you measure the speed of light, the result would be different. The slower clock would measure a faster speed of light, and the faster clock will register a slower speed of light. Thus, every scientific principle or law that we formulate would turn out to be useless. Given this problem, we are compelled to assume the uniformity of space not merely for the reasons of the conservation of momentum and angular momentum (as we discussed above), but also for the postulate that laws are universal.

This makes the problem slightly worse; it not just that the laws of conservation of energy, momentum, and angular momentum are tautological assumptions, but every law of nature—e.g. Newton's law of gravitation—depends on such tautologies. The laws could in principle be non-uniform, and that non-uniformity could be caught if we had a way of detecting space and time non-uniformities. But since we cannot detect space and time non-uniformities, therefore, we cannot know if the natural laws fail. Indeed, quite so often, we use the natural laws to determine length and time. For example, we can define a unit of length as the distance covered by a standard falling mass under the laws of gravitation on earth. Now, time and mass are standard properties, whereas length is a derived property. Likewise, if we use electromagnetism to compute time—e.g. define a unit of time by the duration taken by 100 cycles of an electromagnetic wave of a standard wavelength (defined by the length derived from mass)—then we can replace time and distance by mass and charge as the definitions of all other standards.

The conclusion is simple: we need three properties—e.g. time, length, and angle—but these three could be anything: we can substitute length by mass and substitute time by charge. The tautological properties of conservation laws (which we noted earlier) can thus be extended to all other laws and properties. If we can define some mathematical properties that have definite relations to length, time, and angle, and we have a way of measuring these properties, then they can easily replace length, time, and angle. It doesn't matter what these properties are. They just need to be measurable by some standard, and have well-defined relationships to time, length, and angle. Thus, every physical property law depends on space and time.

We can illustrate this fact by the example of the speed of light. For instance, you can say: The uniformity of space and time is proven by the fact that the speed of light is measured to be a constant at different places. But the speed of light is a *ratio* of distance and duration. If both space and time were not uniform—e.g. some clocks ran slower as the angles on the dial were larger—then, the ratio of distance to time would still be constant. If you counterargue: The sum of these angles must be 3600, so the angles cannot be changed forever, the counterargument is that you are assuming a flat space. (You can visualize a twisted clock by imagining a picture of a clock drawn on crumpled paper.) If space were not flat, then angles don't have to add up to 3600 and due to spatial elongation, the clock will also run slowly, and we would not know if time and space are indeed uniform.

The point is this: whatever we consider to be a *proven* law of nature in a flat and uniform space and time, may also be proven in a warped and nonuniform space and time, with the warping and nonuniformity adjusted in unique ways. If space and time don't warp consistently to *hide* the ensuing differences, we would not attribute the differences in outcomes to the failure of a law. Rather, we will attribute that to the law itself not being universal, because the universality assumes a uniform space and time. Hence, if space and time were not uniform, then genuine failures of laws attributable to the nonuniformity of space and time would be attributed to the failures of the laws. Likewise, genuine failures of laws attributable to the laws, would be hidden by the compensatory nonuniformity of space and time. In short, we just don't know how to interpret empirical variations in the data.

For example, variations in the speed of light have been observed[2,3], but physics assumes that the constant speed of light is a law of nature. Now, should we say that the law is universal, and blame the observed variations in speed to errors in experimentation? Or should we say that the law is itself not true? Or should we say that the law is true, and the experimental variation is true, but the space and time are not uniform and isotropic? You can see how this problem wreaks havoc on everything known in science since every 'proof' can be doubted, and every 'refutation' can be attributed either to minor experimental errors, or to the variations in space and time, or to the failure of the laws itself. Since we cannot know which of these attributes is truly the cause of the observed failure of a law, or the observed validity of the law, therefore, we cannot know if the time passes uniformly.

The Oscillator Hypothesis

There is a solution to this problem, and it requires us to change our notion of space—locations in space must now comprise small *oscillators* that vibrate in different directions—X, Y, and Z^4. I will call this *The Oscillator Hypothesis*. Recall that the problem of spatial uniformity requires us to step outside space, such that different locations in space become objects to be measured by other objects. Then, we can treat space just like we treat material objects. The Oscillator Hypothesis achieves this. It says: each location in space is an object, such that we can measure these objects against one another. The property of each oscillator is a frequency and a form of vibration. If two oscillators have the same frequency, then they would have the same measure of time. Similarly, if two oscillators have the same form of vibration, then we can say that they encode the same information.

The 'observation' of the world can now be defined as the ability to synchronize the oscillators with each other. For instance, if an oscillator that represents the observer's *senses* were synchronized with the oscillator that represents a measured *object*, then, by synchronization, the observer's senses would vibrate in the same way as the measured object, and that similarity in the oscillations would indicate that the senses have the same information as the object they are measuring, and similarity now equal to 'knowledge'—i.e. the sense has acquired the property of the object it was meant to know.

Now, if time in the measured object changes, then so will its vibrational frequency and this change will make the senses unsynchronized with the object. To keep the senses synchronized with the object, the senses would have to change, and by that change we can know that the time in the object is now running differently. Similar types of synchronizations can also be construed for the form of the oscillation, and the direction of oscillation. If the senses and the objects are synchronized, then the sense 'knows' the world. If the world changes, then the senses also change, and that change in the senses can be understood as the change in the world. Thus, if each location in space is defined as an oscillator, then we can talk about whether two observers, or objects, or an observer and an object have the same measure of time, distance, and direction. If some law of nature is confirmed in different places, then by the fact that they are mutually synchronized, we can say that the law was indeed confirmed using the same *standards*.

To achieve this, an oscillator must have three key properties: (1) a frequency of vibration, (2) a form of vibration, and a (3) direction of vibration. These properties are distinct classical conceptions of space and time, but they are combined in the oscillator. As a result, we are not talking about three instruments each to measure distance, direction, and duration. We are talking about a single pattern, which represents all the three properties.

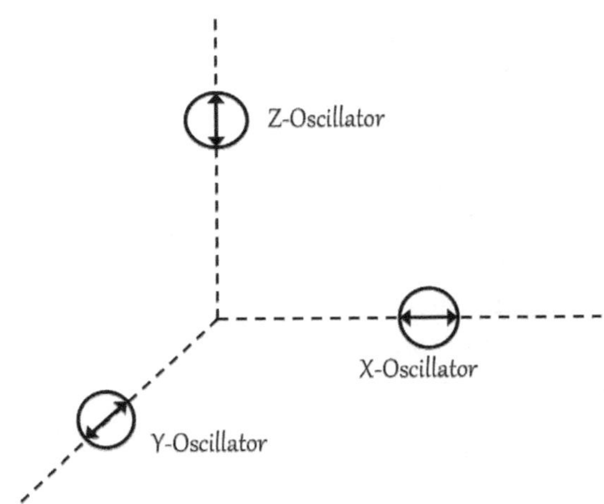

Figure-11 Oscillators Define Three Properties

The new problem is this: How do we synchronize the oscillators? Obviously, synchronization requires some exchange of information, but that exchange necessitates the existence of some 'space' between the oscillators. If the oscillators are location in space, then there is nothing between them. But if there is no 'space' between these oscillators then how will we exchange information to even know if the oscillators are synchronized?

This problem doesn't arise in classical physics because space is comprised of a continuous sequence of points, and due to this continuity, there is nothing 'outside' space. Distance between two points can be, in principle, obtained by measuring the number of points between any two points. However, in practice, this idea turns out to be flawed, because on a continuous line, the total number of points between any two points in *infinite*[5]. In fact, real number arithmetic shows that the number of points between any two points is *equally infinite*. As a result, we cannot say that distance between points 1 and 2 is half the distance between the points 2 and 4. Since the number of points between any two points is identical, therefore, we can squeeze the distance between 2 and 4 and make it equal to the distance between 1 and 2, and no point would 'fall out' of the line—because the number of points is equally infinite. In effect, you can squeeze the space of the entire universe to a ball that fits in the palm of your hand, and it will still contain as many points as it did before. The ability to squeeze the distance without losing points underlies the fundamental issue of distance measurement: we cannot know what the real length is, so, even if the length were contracted or expanded, we have no way of knowing. Now, distance is defined not as an inherent property of space, but by an *object in space*—e.g. a meter. The belief is that matter cannot be squeezed like space. If we try to squeeze matter, then some material particle will fall out of the meter, and then it will also be shorter. This idea, however, doesn't stand on solid footing, because in classical physics, a meter is also comprised of infinitesimal particles. If the space could be squeezed indefinitely, then it should also be possible to squeeze a meter indefinitely. But we know that this isn't the case, so we assume that a meter cannot be squeezed, and we use it as a measure of distance, but distance is a property of the meter, and not of space.

Therefore, while it is true that the problem of distance measurement doesn't arise in classical physics, it is not because the definition of space is better; the classical physical definition of space is also problematic, but the problem is solved by *assuming* continuity, attributing length to the meter, and thereby indirectly attributing length to space (since the meter measures some space). The trouble is this: How do we know that the distance between the markings 1 and 2 on the meter is twice the distance between the markings 2 and 4? The short answer is that the mass of the scale between 2 and 4 would be twice the mass of the scale between 1 and 2. But how can we measure the mass—if not by using some measure of length? And the answer is that we cannot. The only real measure of space is if we can count the atoms in the meter, because then we can say that there are twice as many atoms in the meter from the marking 2 to 4 as there are from the marking between 1 and 2. This would be in lieu of our ability to count the number of points between any two points. The assumption inherent in this counting is that each atom has a finite size and is not an infinitesimal point, because if it were infinitesimal then there would be an equal number of atoms between any two points, and it would be impossible to measure distance.

Therefore, when we treat space itself as a collection of discrete oscillators, we are only pushing the counting of atoms in the meter to the counting of similarly atomic locations in space. The foundation upon which we measure length using a meter is also the foundation on which we can measure distance in space, and that foundation is that there cannot be an infinite number of infinitesimal points between any two points. When each location is an oscillator, then the points cannot be infinitesimal. And if these points are not infinitesimal, then there cannot be infinitely many of them.

Hierarchical Space

Now, we are back to the question: How do we synchronize the oscillators, to know the duration, direction, and distance at another location? One possible answer to this problem is that these oscillators must be organized *hierarchically* such that the 'higher-level' oscillator is the 'space' between the 'lower-level' oscillators. In short, the lower-level oscillator is a *part* of the higher-level oscillator, such that the higher-level oscillator is the *whole*.

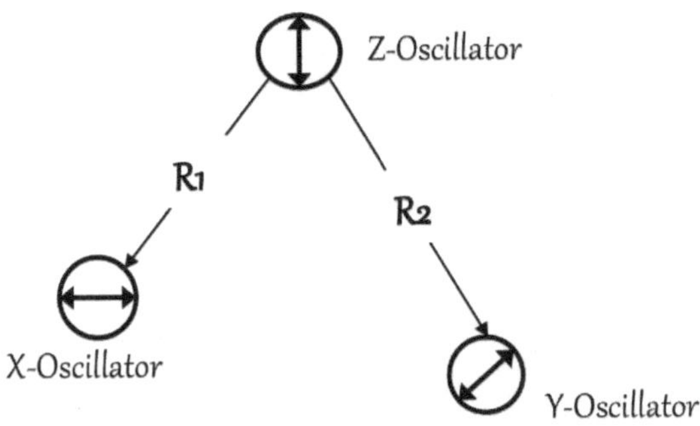

Figure-12 Hierarchical Relation Between Oscillators

The distance between two lower-level oscillators can be measured if we know the *relation* between the higher- and lower-level oscillators. In

Figure-12 we depict this idea by the relations R1 and R2. For the sake of illustration, we show the X- and Y-Oscillators as being physically apart from the Z-Oscillator. But this is only for illustration. If these entities were indeed apart, then there would be no way to measure the distance between X, Y, and Z, because there is nothing in between them. The distance measurement requires an underlying space, and that space is the Z-Oscillator. It is a *location* in space, but it is also *space*. When we think of the Z-Oscillator as a location in space, then we are speaking of Z relative to an even higher-level oscillator which constitutes the space in which Z is itself located. But when we describe Z as space, then it is in relation to the X- and Y- locations. A more accurate depiction of these locations is thus Figure-13 where the lower-level locations (X and Y) are seen as a part of the higher-level location (Z).

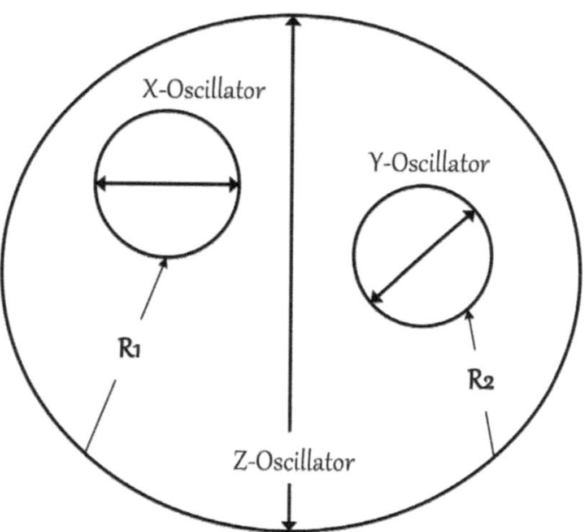

Figure-13 Containment Relation Between Oscillators

Such hierarchical constructions are common in our everyday world. For example, if you ask a person about to board an international flight from France to India: "Where are you going?" he might say: "India". Once he lands in India, and is about to board another domestic flight, the answer to the same question—i.e. "Where are you going?"—could be "New Delhi". Space in our everyday intuitions is organized hierarchically from wholes to parts. When we send a postal mail, the address is described hierarchically (in the inverse order)—a house number, a street name, a city, a state, and a country. Likewise, while sending emails, the addresses are defined hierarchically from a person's name, to an organization's name, to a top-level domain. Internet addresses are also assigned hierarchically in the same way.

This method of addressing locations in space is not used in classical physics; we don't count top-down from whole to parts; we rather count in *three* directions—e.g. X, Y, and Z. Conversely, in the hierarchical space, we always count from top to down, from whole to parts. Indeed, because the locations are discrete, therefore, all these locations can be numbered 1, 2, 3, etc. which means that they can 'fit within' a single dimension. Since there is no fundamental difference between a dimension and a location on that dimension—i.e. they are both locations, if we look at space from the highest point in the space—therefore, the highest location constitutes the 'dimension' and every other location becomes a point along this dimension.

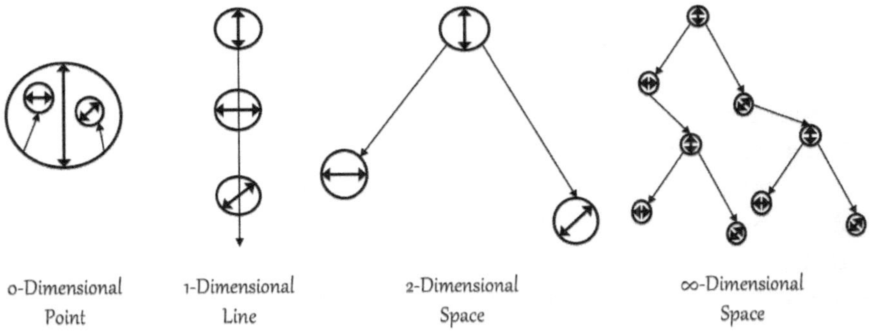

Figure-14 Different Descriptions of Hierarchical Space

As a result, we can say that the universe is zero-dimensional—because the entire universe is present inside the highest location. We can say that the universe is one-dimensional—because the highest location is a dimension and all other dimensions can be locations on the dimension represented by the highest dimension. We can say that the universe is three-dimensional because the inverted tree like structure requires three distinct types. And we can say that the universe is infinite-dimensional because each location is after all a dimension—from the vantage point of the lowest location.

The Process of Measurement

Once we have redefined the nature of space hierarchically, then we can return to the problem of oscillator synchronization, by which we can measure the properties of distance, duration, and angle. We must still solve the problem of classical interactions, which involve an energy transfer. Classical physics postulates that we can measure the state of a system without altering the system itself, but there is no conceptual basis for this idea. For example, if you were measuring the temperature of an object, then the thermometer would absorb some energy, thereby lowering the temperature of the measured object, and the state being measured would not be identical to the state that existed prior to measurement. All measurements are lossy, but this issue is ignored in classical physics because the losses are small compared to the properties being measured; after all, we are measuring macroscopic objects, and the energy required during the measurement of a classical property is small relative to the properties in the objects. This problem, however, rears its head when the properties being measured are no longer small compared to the loss in measurement; for example, measurements on atomic objects cannot claim to be measuring the properties of the world as it were prior to measurement. If during a measurement, an oscillator loses or gains energy, then the measurement would be inaccurate. In transferring the energy during a measurement, the measured object would change its state and we can no longer say that we are measuring reality as it were prior to measurement. Likewise, if the observer's state is altered in a measurement, then how can the observer compare the object's state with its own state to know if these two states (indicating duration, direction, and distance) are the same?

In classical measurements, the properties of atomic particles are *inferred* from the effects. For example, to measure the structure of a crystal, the crystal is irradiated with light, and four kinds of effects are measured—(1) some light is absorbed, (2) some light is reflected, (3) some light passes through, and (4) some light is refracted. Under the classical physical model of measurement, we would suppose that the measurement of the crystal corresponds to the reflected light, but that is not true of quantum particles, where all the above effects must be incorporated into a measurement. Moreover, due to quantum entanglement (which I will discuss in a subsequent chapter), each reflector and refractor must be treated as a unique source of light, and all these sources must interfere with each other producing a complex interference pattern. All of these are the *side effects* of the irradiation of the crystal, and the crystal structure must be *inferred* from the interference pattern produced by the interference of millions of irradiated atoms[6].

Does this mean that we never truly know the nature of the world as it exists prior to our knowing? That would certainly be the case if measurements were treated classically. However, this would not be the case, if we had the ability to measure the world without *losing* or *gaining* energy.

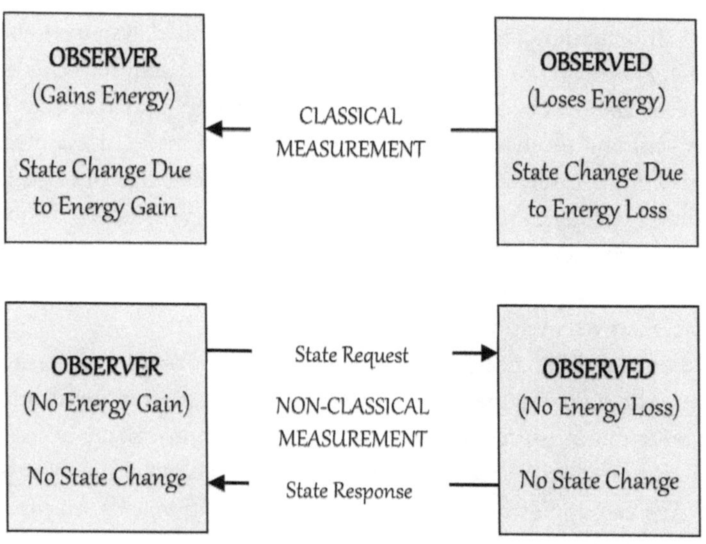

Figure-15 Classical vs. Non-Classical Measurements

One possible method of such measurement is if the state of the object is *copied* from the object and transferred to the receiver *as a copy*. When a receiver gets this information from the source, it stores the information as a copy. The information can now be compared with the observer's state, without changing the observer's state, or the state of the measured object. In this non-classical method of measurement, energy is not lost by the measured object, but some energy is gained by the measuring system. For energy to be conserved, the observer must emit some energy toward the measured system, which will then copy the state of the measured object and become a *symbol* of the measured system's state. When this symbol returns to the observer, information about the measured system is gleaned, but in the process, neither side gains or loses any energy (the energy emitted by the observer is returned to the observer after the measurement is done). Figure-15 shows the difference between the two kinds of measurements.

In *physical* measurements, energy flows from the observed to the observer, and this changes the states of the measured and measuring systems. However, computers measure the system state in a different way: a control system sends a *request message* to the controlled system, asking for information about its state, and the controlled system then sends a *response message* containing the requested information. This model of measurement is based on information exchange and the quantity of energy sent and received does not indicate the properties of the measured system. For example, the request and response messages could be longer or shorter, and hence require greater or lesser energy, however, that would not change the actual properties in the measured system. Therefore, there is no correlation between the actual properties of the measured system and the energy sent and received.

The crucial difference between physical and informational methods of knowing the state of a measured system is that a receiver requests information about the measured system, and then gets information in response. If this request were not issued, then the response would not be received either. We can apply this idea to conscious perception. We see the world if our senses 'request' information from an object and receive a 'response' from the object. If the senses do not issue a 'request' and consequently don't receive a 'response' then we would not be able to perceive the world. For example, my eyes could be open, and capable of seeing, but if the eyes do not send a 'request' and receive a 'response', then they will not see. The conclusion would be that we can know the world but only if *want* to know, and only if we ask the right *questions* (i.e. send the correct requests)[7].

This model of perception is attractive because it helps us solve the problem of oscillator synchronization—e.g. observation—without gain and loss of energy. Under this model, each oscillator must issue a 'request' and receive a 'response', such that the response indicates the state of the measured object, by *representing* the object's state as symbolic information. This constitutes a non-classical and a non-physical model of observation. With this model of perception, one oscillator can determine the state of another oscillator, without changing either oscillator, and this is very important because we are trying to determine if the two oscillators have the same sense of duration, direction, and distance. Now, it is possible to answer the question: "Does time pass uniformly?" because we have a method to know.

So, Does Time Pass Uniformly?

I apologize if this has been long-winded. We had to establish that we can know distance, direction, and duration, before we say whether these are uniform. We saw that this is impossible when space and time are continuous containers. The hierarchical notion of space makes it possible to determine the answers. But this solution also tells us that distance, direction, and duration would *not* be uniform, as these properties are defined relatively.

For example, the meaning of "10th Street" is not obvious unless we describe the town or city. There are indeed many "London" and "Paris" in this world because they are parts of different states and countries. In India, practically every city has a "Mahatma Gandhi Road". Similarly, in a hierarchical internet address, each number has a meaning and reference only in relation to the previous number. You might say: "I will meet you at 2 o'clock" but that is relative to your time zone. Or, you can say that "to reach this place go East", but that is only relative to where you are right now.

We can now say: this problem of relativism can be solved if we know the complete hierarchy in relation to which the properties are defined. *If* we knew the hierarchy, will we conclude that distances, directions, and durations are uniform? Of course, we can know the properties accurately if we know the complete hierarchy. However, that won't solve any problem, because by knowing the absolute values, people will still *report* different facts about the same world. For example, if there are places along the Sun's orbit, and we consider an absolute viewpoint, then places to the South would say that the "Sun rises in the East", while those on the North would say that the "Sun rises in the West"—and that clearly shows how using an absolute viewpoint about space and time doesn't lead to similarity of experience.

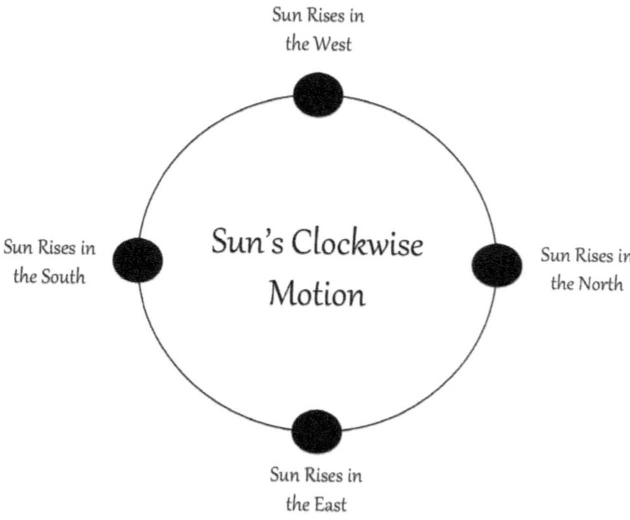

Figure-16 Relativized Perception of the World

It is a fact of life that we have chosen 'meter' and 'foot' as measures of distance because our bodies are similarly sized. If our bodies were bigger or smaller, then we would use other kinds of standards. Likewise, if the durations of our lives were different, then the measures of time would be different. The converse is also true—i.e. if the measure of length is different, then bodies will be differently sized. And if the measure of time is different, then durations of our lives will also be different. Thus, every fact of life must be defined relative to the space and time we inhabit, and we might wonder: Is it better to have different standards or different values? For example, should we say that all of us live for 100 years, if 'year' is a relative term? Should we say that the Sun always rises in the East, but

East is a relative term? Should we say that the average height of people is 1.5 meter, but 'meter' is a relative term? While communicating with an observer in a remote location, should we use relative language (i.e. identical words with relativized meanings), or absolute language (i.e. relativized words with identical meanings)?

If we use an absolute convention of words and meanings, then we will frame our descriptions about the same thing, using different words. For example, one observer will say that the Sun rises in the East, while another one will say that the Sun rises in the West. To understand these descriptions, we will always need to keep a dictionary handy to determine that East means West, and vice versa. In some sense, therefore, it is preferable to use a relativized language so that we can agree upon the description of reality across the places and times, although what we observe can be quite different. This is a condition imposed by the need for communication between places. And once we decide to communicate, then we must still use the same words, but the meanings of these words would not be identical across the places.

The conclusion is that distance, duration, and direction are not uniform throughout space and time—if we use the same language. However, by using the same language, the universe *seems* to be uniform, because our requests and responses about information from that place would be encoded in the same language, but the words mean differently: the same words are used but the meanings of those words are relativized to that place and time.

We will discuss the problems arising out of relativity in the next chapter, but we can quickly contrast *physical* and *semantic* relativities here. Physical relativity entails that different relative space-times have the same experience and verify the same laws, but each observer thinks that the other observers are having a different experience—because we think we are directly observing them. Semantic relativity entails that different relative space-times have different experiences and don't verify the same laws, but we can think that the other observers have the same experience and verify the same laws—since we are getting their *reports* in which the words mean differently.

The crucial difference between physical and semantic relativity arises by the change in the definition of a *measurement*: Are we measuring the world physically—as if we were truly there? Or are we receiving *information* about the world, that is encoded in some communicable universal language? Physical relativity assumes that we are measuring the world as if we were truly there, performing a *local measurement*. But this is a false claim. Even when we are in proximity with the world, we are only sending requests and receiving responses, and this is obvious to anyone who has tried sense control: the existence of the ability to perceive doesn't mean that you must perceive. It is an *ability*, which means that it can be used selectively and by will. And when our perceptive process is understood, then all measurements become semantic exchanges and the role of language manifests clearly in science. To observe the world, we must now say that there is a *shared natural language* of words and meanings, and information comes to us through this language—even when we perform a measurement. The words of this language, however, don't mean the same to everyone, at every place, time, or direction. The commonality of words and the variation of their meanings entails that the use of a common language does not entail a universality of facts.

As a result, we can say that 'that place' is just like 'this place', but if the same person went to that place (and carried the memories of 'this place') he will realize that the two places are different, but it was impossible to know that until we changed the places. Why? Because when you go to the new place, you can continue to use the same words, but every word will now have a different meaning. That change of meaning of the words constitutes the difference between *being there* and knowing that place from *here*.

Thus, physical relativity is true in a *linguistic sense* that if we exchange information with another place about that place, then we will believe that the same laws hold true, and the experiences are the same, although it is false in the *real sense*. Conversely, semantic relativity is true in a *real sense* that all places have different laws and experiences, but false in a *linguistic sense* that identical reports of confirmed laws and facts are not actually true. You might argue: But we have empirically confirmed that physical relativity is true! Doesn't empirical confirmation prove the truth? For example, relativity entails that there is no absolute space and time, and that the universe is uniform in its properties and its laws. The short answer is that all these claims are false, but they *seem* to be true because of linguistic relativity. We are receiving linguistically encoded reports that *tell us* that the world is the same everywhere, but the meanings of these words are different—there and here. The analysis of the report leads to the false conclusion that the universe is uniform, but real experience of that place will demonstrate otherwise.

Takeaways from Non-Uniformity

The discussion about whether space and time are uniform, and the answer to these questions leads us to some key ideas that will prove useful later in the book. We can recapitulate these problems and solutions:

- Classical physics postulates that space and time are uniform and isotropic, and these assumptions lead to 'laws' of conservation of energy, momentum, and angular momentum. But it is impossible to verify these laws because we must step outside space and time to know if they are uniform and isotropic. From within space and time, it is impossible to know if time moves uniformly or not.

- This problem can be solved if space is described as discrete locations comprising oscillators whose vibrational direction indicates a direction. However, discreteness of locations leads to a new problem: How do we measure the *distance* between these locations?

- The solution to this problem is that locations must be described *hierarchically*, or as the relation between wholes and parts, and the distance between two lower-level locations is given within the higher-level location, and thus we get both distance and direction.

- We are now led to the problem of *measuring* the direction and duration at the different locations, and the physical model of measurement is flawed because it involves an energy transfer which will then disturb the states of both measuring and measured systems, and the result would be that we will never know the real direction and duration at the various locations in space. We address this problem by defining an informational model of observation that employs a linguistic encoding of values in terms of symbols.

- Obviously, to transact information, we must use a common

language, but the hierarchical view of space—which was necessary to even measure distance, duration, and direction—forbids shared meanings. In short, in the physical view, we can never know if the universe is uniform, and we must assume it. In the semantic view, we can know that the universe is non-uniform *if* we go to different parts of the universe; from the present location, we cannot know the non-uniformity, and linguistic reports will show uniformity.

- Thus, space and time are not uniform across the universe, and assumptions to that effect in science are false. Even the empirical confirmations of such uniformity are false due to linguistic relativity. We can go to those places and find the non-uniformity. However, the non-uniformity of space means that we cannot go to those places with the same bodily, sensual, and mental apparatus.

- From the previous chapter we can recall that possibilities change as an effect and consequence of responsible or irresponsible action. Therefore, if space is described as the domain of possibilities, then going to another place would involve changing the possibilities of experience, and that would be an effect of responsible or irresponsible choices. Thus, we cannot go to another place in the same bodily, sensual, and mental apparatus, but the apparatus must change as a result of choices, and that change would transport us to a different part of the possibility space—which is 'higher' or 'lower'—because the possibility space is now also a hierarchical space.

- The hierarchy of space now has a semantic interpretation: the same words can be used in all parts of the universe, but the experiences corresponding to these words would not be the same. The words of communication can be universal, but we cannot *experience* the meanings of those words in the present body, sense, and mind.

5

Is Time Absolute or Relative?

The Theory of Relativity

The theory of relativity (there are two theories of relativity—one *special* and the other *general*; in this chapter, when I use the term 'relativity' I mean the theory of special relativity; I will discuss the theory of general relativity in the last chapter of this book) emerged out of Einstein's efforts to reconcile the constant speed of light with the relativity of speed of ordinary objects. If a ball is moving toward you, as you move toward the ball, then the speed at which the ball hits you would be the sum of your speed and the ball's speed. Conversely, if you were moving away from the ball, then the speed at which the ball hits you will be your speed minus the ball's speed (if you are moving away from the ball faster than the ball's speed, then the ball will never hit you). This fact about ordinary objects proves false in the case of light, as it is seen that all observers measure a constant speed of light.

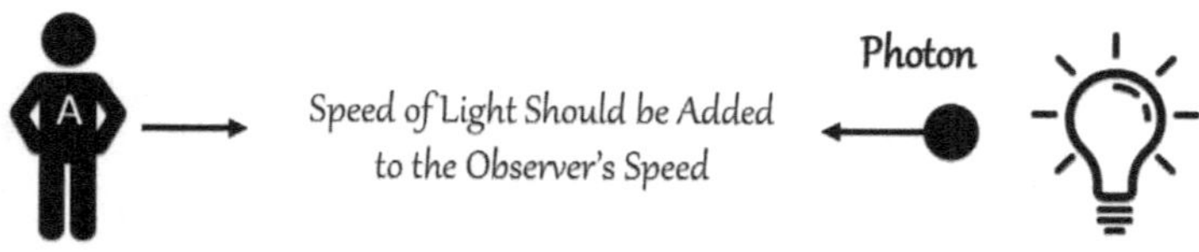

Constant Speed of Light is Measured by Both Observers

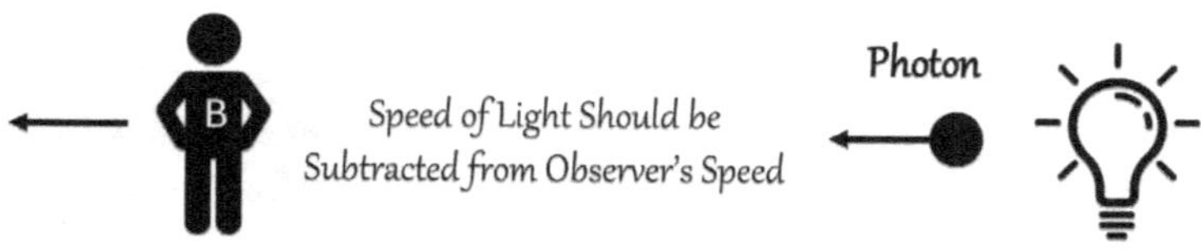

Figure-17 The Observed Constant Speed of Light

Moreover, if a third observer were observing the above situation, then he or she will observe that light indeed reaches the destination *earlier* for the first observer, since he is moving toward the light source, and *later* for the second observer, since he is moving away from the light source. Therefore, even the third observer measures the same (identical) speed of light.

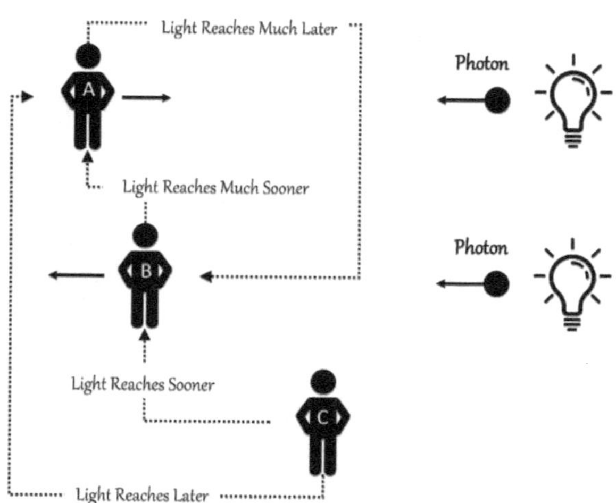

Figure-18 Four Viewpoints of the Same Event

Thus, there are four views of the events: (1) C observes that the light arrives at B earlier than B's detection, (2) C observes that the light arrives at A later than A's detection, (3) A observes that B receives light much later than B's detection, and (4) B observes that A receives light much earlier than   A's detection. Thus, everyone measures the constant speed of light, but everyone also attributes a different time for the same event. Hence, the constant speed of light must be treated as a *law* of nature. Recall from the earlier discussion that for time to pass, there must be a strict ordering of events—i.e. a sense of 'before' and 'after'—after which we can apply the idea of 'past' and 'future'. Relativity disturbs the strict ordering for different observers because the same event can be seen to occur before or after, for different observers. When the strict ordering is disturbed, then the notions of past, present, and future are also different for the different observers. Thus, something that is in the future for one observer may be in the past for another observer, and vice versa. Thus, the words 'space' and 'time' lose their classical *objectivity*—(1) the strict ordering of events doesn't exist, (2) the notion of past, present, and future changes, and (3) the durations and distances between the events become dependent upon the observer's motion.

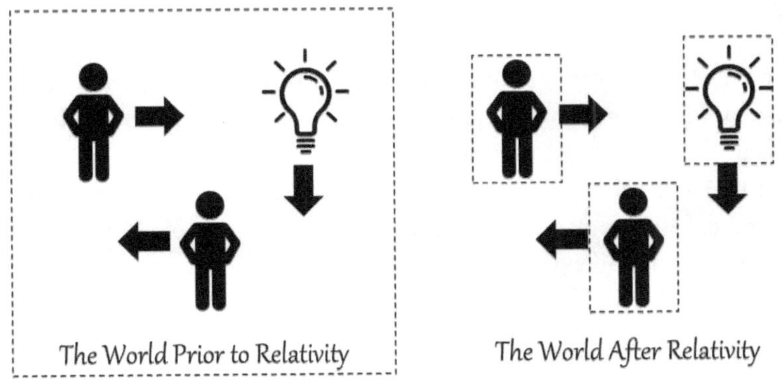

Figure-19 The World Before and After Relativity

Two such effects of relativity are known as *length contraction* and *time dilation*. Both pertain to the observations from a moving frame. The measurement in the static frame measures the maximum length, but the lengths contract in the moving frame. Likewise, the measurement in the static frame measures the shortest time, but times dilate in the moving frame.

Due to the loss of space and time objectivity, physicists say that space and time are not *absolute* containers of the world. Rather, each observer carries their own space and time, and hence, space and time are *relative* to the observer. In short, the world is not *in* space and time (as was believed to be the case prior to advent of relativity); rather, each object has its own space and time, what the world is *in* is no longer amenable to scientific inquiry.

The Objectivity of Physical Properties

As we have discussed, all physical properties are measured using distance and duration. Therefore, if lengths contract and times dilate for a moving frame, does that mean that these observers will also measure varying values of physical properties? For example, temperature is measured by the distance by which the mercury moves in a thermometer. If lengths contract due to relativity, then what happens to the temperature? Does mercury expand less along with length contraction such that the observed temperature remains the same? Or does mercury expand the same such that the temperature is seen to rise? Similarly, a weighing scale measures mass by the movement of a pointer. If length contracts, then what happens to the mass? Does gravity reduce along with the length contraction, such that the value of the mass remains the same? Or does gravity stay the same, but because the length has contracted, therefore, we find that the mass has increased?

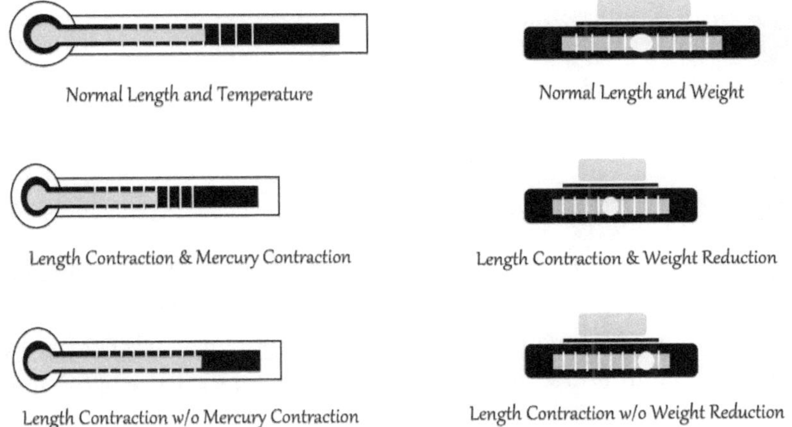

Figure-20 Possible Effects of Relativity on Physical Properties

Remember that if mass reduced due to length contraction, or if the temperature reduced to due to length contraction, then the principle of energy conservation would be violated. Hence, even if the distances and durations change, there can be no change to the mass or temperature of objects. However, since distances and durations are indeed changing due to relativistic motion, therefore, we must obtain greater or lesser *values* for these properties, even though there should not be a real change to energy.

Hence, one person would say that an object is hotter while the other will say that it is cooler. One person will say that the object is heavier, while the other will say it is lighter. Relativity describes this by saying that in its own reference frame, a particle has a *rest mass*. But in other frames, where it moves, it also has a *relativistic mass*. If the particle is stationary in a frame, then there is no length contraction and the rest mass is the 'true' mass. But if the particle is moving in a frame, then with length contraction the relativistic mass is higher by the factor that the length contracts. In summary:

$$m_{rel} = \frac{1}{\sqrt{1 - \frac{v^2}{c^2}}}$$

$$t_{rel} = \frac{t}{\sqrt{1 - \frac{v^2}{c^2}}}$$

$$l_{rel} = l\sqrt{1 - \frac{v^2}{c^2}}$$

Note how as the length contracts, the mass increases by the same factor as the contraction. If the length becomes half, then the mass doubles. This is keeping in view that the *scale* of measuring mass has contracted, therefore, the measured value (employing this contracted length) must double.

Now, we are led to a drastic conclusion, namely, that if the values of temperature and mass are observed to change, without a real change in these properties, why should we not treat the change to the values of length and time any different? Why would we not say that length contraction and time dilation are simply *observed phenomena* and not a change to the reality? If the observed values were indeed the reality, then as length contracts, every other property must also contract with it. For example, the mercury must expand lesser in the thermometer, to indicate the same temperature. Likewise, the pointer on the weighing scale must move lesser, to indicate the same weight. Since this is not true, therefore, length contraction and time dilation are only changes to perceived values, rather than changes to reality. And this means that there is an objective reality 'out there' and everything *seems* to have changed, but if we take this seeming change to be real, then the fundamental principles of nature, such as energy conservation, would be broken.

When relativity was formulated, modern science was gripped by *positivism*[1] or the idea that the job of science was not to speculate about the nature of reality, but only to explain the observations. Based on this idea of science, all questions of realism about space and time were abandoned. But that rejection of realism isn't enough, unless we reject all other kinds of realism—e.g. that there is no objective mass, no objective temperature, and no objective energy. Obviously, the positivist stance must extend to every physical property—if it is to be accepted as true—but nobody is prepared to accept that mass, temperature, and energy are not objective, because it is upon this objectivity that the realism of science rests. If mass and energy are not objective, then their conservation must be false. But if we accept the objectivity of mass and energy, then why should we not apply it to space and time as well? Why should positivism be applied selectively and not universally? The conclusion is that because energy, mass, and temperature are indeed objective, therefore, we must also reject the relativism of space and time.

The new question is: If the values of duration and distance are attributed to the observer, while there is an external objective reality, then this reality must also exist in some objective and real space and time, which is different from the observed space and time. In short, the objectivity of energy, mass, and temperature leads us to the notion that there must be an *absolute* space and time, and the relativity of distance and duration creates a *relative* space and time, but these are perceptual or phenomenal space and time. The problem of relativity is the existence of two such domains.

The Twin Paradox

A similar problem arises in the case of the *Twin Paradox*[2] which seems to make the "age" of a person a subjective construct. If one twin leaves the earth in a spaceship and returns to earth after some years, then he or she would find that the twin that stayed on earth has aged more. But aging is not simply about a clock having more rotations. With aging, our body also becomes weaker, the metabolism slows down, and the immunity to diseases generally declines. Therefore, there is a real material change to our body with aging, rather than merely the clock. However, as we have noted above, the material properties of the object do not change as the effect of relativity, because that would violate the principles of energy conservation. The objectivity of these properties governs aging, not merely the clock rotation. And since these properties remain the same for the twins, therefore, each of the twins would factually be of the same age—in terms of their bodily strength, metabolism, and immunity—although the clocks will differ.

According to relativity, each person will think that the other's clock is moving faster, but that can't be true, so on matter of principle, the Twin Paradox is *true*, and relativity is false. Then, on matter of fact, both twins will have the same age, therefore, the clocks do not determine the age, and so, on the matter of fact, the Twin Paradox is *false*, and hence relativity is false. Thus, relativity becomes false both because the Twin Paradox is true (both clocks run at the same rate) and false (one twin doesn't age more than the other). Relativity can be true if one twin ages more, and one clock runs faster. If both clocks are running equally fast and if both twins age the same, then relativity is false since relativity requires one twin to age more.

Hence, the observed distances and durations are 'illusions' as opposed to 'reality'. But because these illusions will be observed in all experiments, they are akin to our vision of a rail track converging over a distance—everyone sees the same illusion, and yet it is not true. Thus, repeated experiments will prove our theory, and yet, that proof doesn't indicate the nature of objective reality. We must now find a description of the world in an objective space and time, different from the relative space and time, such that the objective reality is consistent with the observed variations in the perceptions. Relativity worsens the problem of time because now there is a real time and a perceived time, and both these times have the problems of time passing, how the time passes, and whether it passes uniformly. If each of these were separately intractable, then relativity doubles the difficulty because it also introduces the distinction between real and apparent time.

The Explanation of Relativity

To address these problems, I will now turn toward the *explanation* of relativity. By explanation, I mean that the perceptions of length contraction and time dilation are like illusions, and the explanation must describe how the illusion arises. Since the illusion in perception is different from reality, therefore, I will use the notions of *absolute* and *relative* space and time. The absolute space, as we have discussed in previous chapters, is the domain of possibilities. Likewise, the absolute time, as already noted, is the choice on these possibilities, which determines the next event. The absolute space and time are different from the observer, and hence the observer's space and time, which constitute the relative space and time are also different from the absolute space and time. The laws of nature pertain to the relation between possibility and choice—which I described earlier as *responsibility*, that leads to effects and consequences. This causality is based on the interaction between possibilities in space, which are hierarchical, and the interactions between these possibilities are informational rather than physical. The solution I will describe requires delving into the details about the nature of this informational causality, and how it leads to the perceptual space and time.

Information comprises meanings and their instances. We can say that the meaning is semantic while the instance is physical[3]. For example, if you were writing a letter, then you encode symbols of meaning, which are instances of meaning. I will call the physical aspect of the symbol its *power*, and the meaning encoded in the symbol as the *strength*. This choice of words is peculiar, so we can also demystify it by saying that greater *strength* of a message means that greater amount of meaning is encoded in the message, and greater *power* of the message also means that the message is longer.

For example, if you were writing a letter, then you obviously need some number of words, which constitutes the *size* of the letter. Then, there is also the *meaning* in the letter. In principle, a lot of meaning can be encoded using fewer words, and a lot of words may be consumed for a simple meaning. For a given meaning, there is a lower limit on the smallest number of words necessary to encode the message. And for a given number of words, there is an upper limit on the largest meaning that can be encoded within it. Therefore, by greater *strength* I mean a greater amount of meaning, and by greater *power* I mean the greater number of symbols. These are two distinct ways of understanding complexity, which we can also call *semantic* and *physical*.

Greater physical complexity means a longer message. For example, a string of a thousand 1's is physically complex because it requires a thousand characters. But this physical complexity can be reduced to a smaller number of characters by replacing the string of 1's by a program that generates these characters. In information theory, such complexity is called *compressible* because we can reduce the number of characters, and algorithms for compression are common these days (if you use a zip program then you are employing compression). Physical compression relies on the *redundancy* in the sequence of characters, and sentences that use more redundant characters are more compressible. A compressed message, however, requires decompression to restore the original message. Hence, if you send a compressed file to someone, then they must decompress it before reading it. Physical compression requires fewer characters during the *transmission* of the message.

There is, however, another kind of compressibility, which we can call *semantic*, which relies not on the reduction of characters merely during transmission, but the *restatement* of the meaning using fewer words. A classic example of such compression is the use of *formulae*. For example, Newton's second law of motion is stated in English as "In an inertial frame of reference, the vector sum of the forces on an object is equal to the mass of that object times the object's acceleration". But the same meaning can also be presented simply as the formula $F = ma$. A computer can compress the English text by removing spaces in the sentence, but it cannot reduce the English sentence to a formula, because it doesn't understand the sentence's meaning. Semantic compression is possible if we know the meaning.

But what is meaning? In simple terms, we can distinguish physical and semantic by a *structure*—physical structure is *linear* and semantic structure is *hierarchical*. For example, to convert Newton's second law into a formula, it is necessary to see the below hierarchical structure in the sequence of words (to keep the picture clear, I did not try to complete the full sentence).

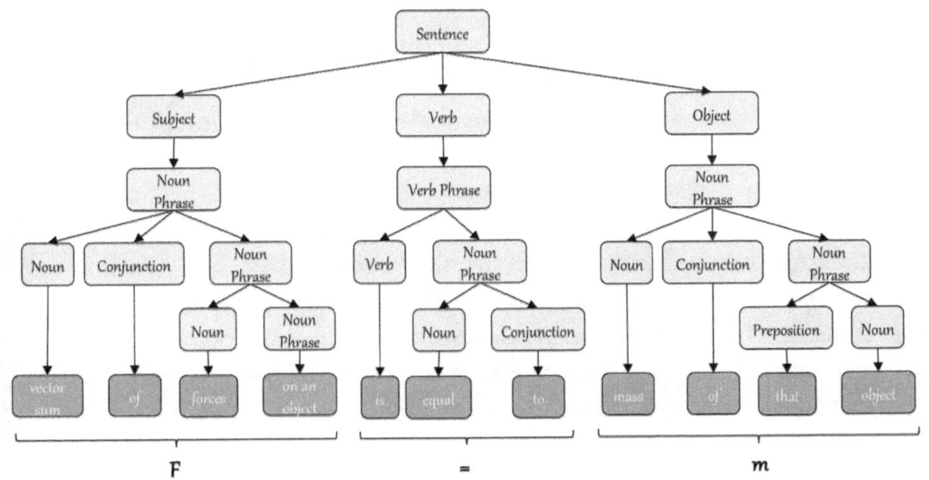

Figure-21 Meanings Emerge from Semantic Hierarchies

Potentially, each sentence can be decoded into many structures. The words can be used alternately as nouns, adjectives, and verbs, which makes it impossible for a computer to interpret the correct structure. The structures change with punctuation, context (of the speaker and listener), by what was said before or after, as well as personality distinctiveness. Moreover, the *references* of the words such as "mass" and "acceleration"—i.e. that they refer to measurement procedures—are outside language. The meaning of a structure, if we consider all the nuances of meaning, and the references in the world, is hence beyond physical sciences and energy-based computers.

Yet, we can talk about semantic information, which comprises a hierarchy of a whole (e.g. a sentence) and its parts (e.g. noun and verb phrases). This information can be encoded and decoded if the information receiver and sender were semantic (rather than physical) systems. Greater complexity in meaning always comes with a steeper hierarchy of the structure. Essentially, the exchange of meaning becomes *stronger* as the amount of *detail* in the message increases. Conversely, the exchange of abstract information requires a smaller hierarchy, and hence a lesser amount of complexity. In simple terms, a bigger tree is more complex and hence a stronger interaction, and a smaller tree is less complex and hence a weaker interaction.

The idea of *power*, in contrast, is simply the number of symbols required to transmit some meaning. As the meaning becomes more complex, a greater *minimum* number of symbols are needed to exchange information. Likewise, as semantic complexity decreases, the minimum number of symbols required for a message reduces. However, since the same number of symbols can be used to transact another meaning, therefore, the relation between meaning and the number of symbols is nuanced—we cannot know the meaning from the number of symbols, but if we know the meaning, then we can determine the minimum number of symbols necessary. In that sense, quality determines quantity, but quantity doesn't determine quality. Of course, quality doesn't *fix* the quantity because we can transact the same semantic information using more than the minimum number of symbols.

Now, we can talk about the nature of interactions between possibilities: they are basically exchange of information, which means the exchange of some meaning encoded through some symbols. But there are some peculiar properties of this exchange that become obvious only when we think of this information exchange as a process occurring during *conscious experience*.

The Peculiarities of Perception

To understand the peculiarities of sense perception, we can examine vision, because it is a model of perception that is easily comprehended. All vision involves two key properties—*contrast* and *brightness*. Contrast pertains to the semantic information and brightness to the physical energy.

The peculiarity of sense perception is that both these properties cannot be high at the same time. Hence, if contrast is increased, then we clearly see more details in a picture, but this clarity is visible only if the brightness is reduced. Thus, higher contrast or greater semantic information comes at the expense of brightness or physical energy. In short, when contrast is increased, then brightness must be decreased[4]. When brightness is decreased, then the picture must also be *close* to the observer, to be perceived. Thus, higher contrast leads to *proximity* or lesser distance. Conversely, higher brightness comes at the expense of contrast or semantic information, and to see such a picture, we must be *far* from it. Thus, higher brightness can be equated to *distance* or lesser proximity. When we look at the 'big picture' then we ignore the details, and therefore, we need lesser contrast and more brightness. Hence, to see the 'big picture', the picture must be *far* from us. Conversely, if we see the 'granular details', then we must also forego the 'big picture'. Then, we need greater contrast and lower brightness, and accordingly, to see the 'granular details', the picture must be *near* the observer.

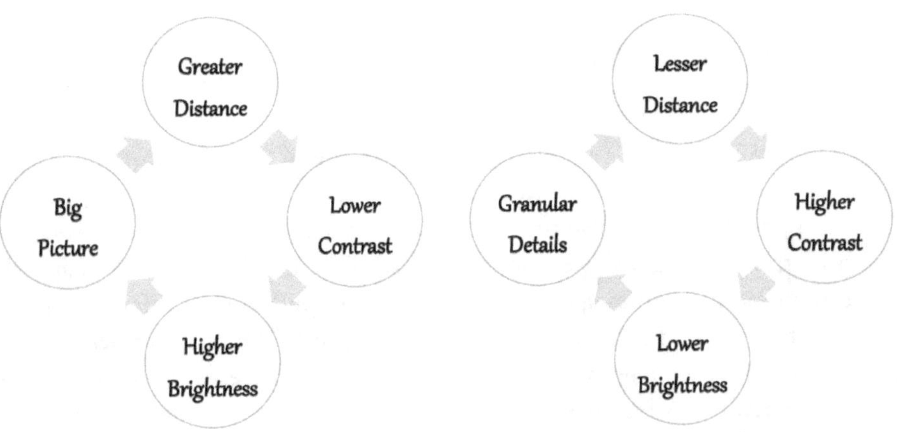

Figure-22 Two Sets of Four Coherent Properties

Thus, we get two coherent and mutually implying sets of properties (in Figure-22). In the first set, the vision of the 'big picture' needs low contrast, high brightness, and greater distance; each property mutually implies the other properties: If contrast is low, then we already know that we are seeing the 'big picture', and which will be farther and brighter. In the second set, the vision of the 'granular details' requires high contrast, low brightness, and lower distance; each of these properties mutually imply each other.

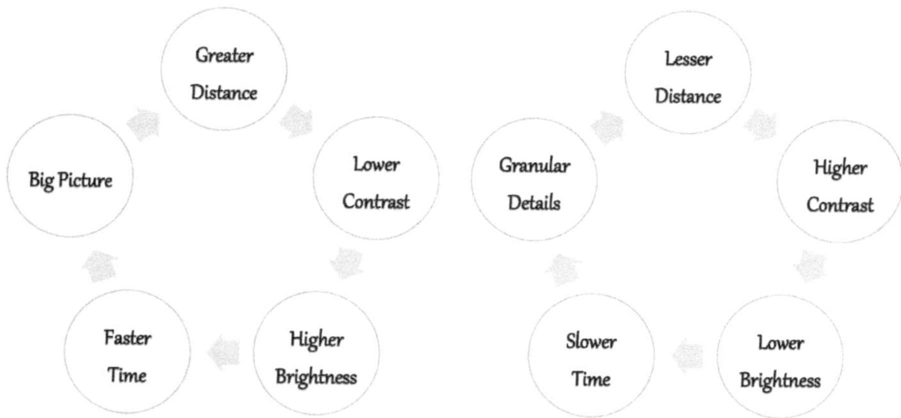

Figure-23 Two Sets of Five Coherent Properties

Higher brightness also implies a faster message transaction, and lower brightness implies a slower message transaction. This is due to the quantum properties of messages (I will discuss the implications of information quantization in greater detail in the next chapter). The main property of all quanta is their energy given by $E = h\nu$ where ν represents the frequency of the quanta. A higher energy quantum has a higher frequency, which implies a lower time for exchanging the energy (frequency and time are inverse properties; when frequency increases, time decreases), and a lower energy quantum has a lower frequency, which requires a higher time. This allows us to update the two sets of properties and add time (given in Figure-23).

Thus, when brightness increases, then the 'big picture' is seen from a distance in a short time. The time required in seeing the big picture is lesser because the amount of semantic information needed for the 'big picture' is lesser. Conversely, when contrast increases, then the 'granular details' are seen from proximity in a longer time. The time required in seeing the granular details is higher because the amount of semantic information needed for the 'granular details' is higher. Thus, we can say that a *perceptual* space and time is created because of semantic interactions in a space of possibilities. When the strength is high, then the power is low, and vice versa. Since the strengths and powers of interactions create a sense of distance and duration, therefore, if these strengths and powers were *changing*, then *motion* would seem to be created, although there is factually no motion. The motion is an epiphenomenon or perceptual illusion of the semantic interactions, and it can be called the *relative* or *phenomenal space*, but it is different from the *absolute space* of possibilities in which the semantic interactions occur.

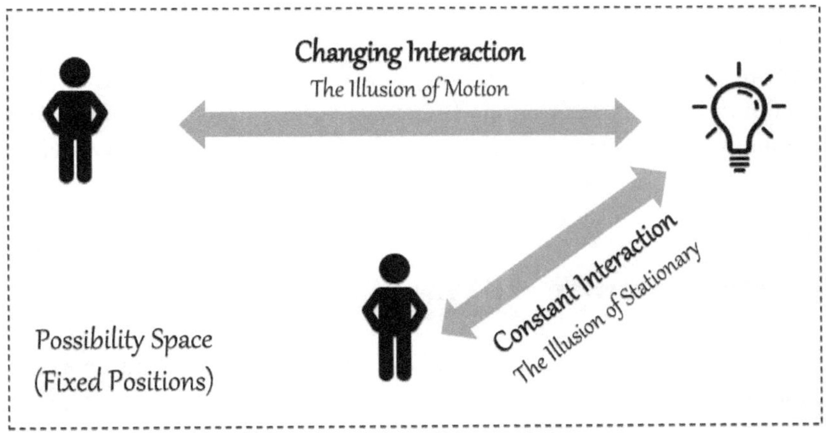

Figure-24 The Genesis of the Illusory Perceptual Space

Thus, 'moving' and 'stationary' are not the real states of the objects. They are both epiphenomena of the semantic interactions. When the strength and power of an interaction is *constant*, then the appearance of stationary is created. Likewise, when the strength and power of the interaction is *varying*, then the appearance of motion is created. In both cases, the possibilities are fixed in an absolute space and hence have fixed positions. Just by the changing strength and power of interactions, therefore, objects can seem to come close or go farther, seem to move, or remain stationary.

With these clarifications about the emergence of stationary and moving states of objects, we can understand why the speed of light is constant. The speed is constant because all objects are stationary. Even when an object seems to move, it is not moving, so its speed of motion need not be added or subtracted from the speed of light. The problem in understanding the constant speed of light arises only if the observers are *moving* relative to each other (or relative to a source of light). If this motion is absent, then the constant speed of light does not present any conceptual difficulties.

The theory of relativity opposes the idea of an absolute space on the grounds that space is the domain in which things *move*. What I have described here is not that kind of absolute space. It is, rather, a space in which things never move. Therefore, there is no conceptual contradiction between the ideas of absolute and relative spaces. But there is a distinction in the ideas of motion: motion is real in relativity, but it is an illusion in the present description. Therefore, we should refrain from a naïve comparison between the rejection of absolute space in relativity and its advocacy here.

Length Contraction and Time Dilation

We can also use these ideas to explain length contraction and time dilation in relativity. To recapitulate, length contraction is the effect that the length of any object will appear foreshortened in the *direction* of motion. Let's quickly clarify a few things. First, if an object moves, this 'motion' is not known to the moving object, because that object always considers itself stationary; the motion is only perceived by another observer (who may not be moving, or moving at a different speed). Second, the observer who perceives this motion considers the moving object foreshortened. Why?

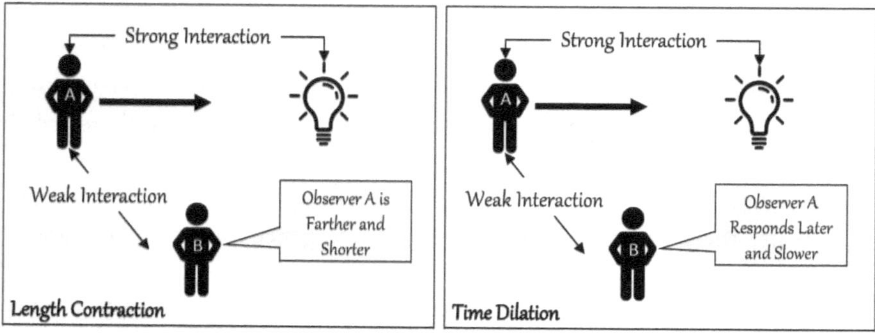

Figure-25 The Informational Explanation of Relativity

When an object moves, it interacts with greater strength to objects or states that are its intended destination. Of course, in classical physics objects don't strictly move toward a destination, although the destination is innate to least action principle formulations of the natural laws. The destination is crucial to address the problems of indeterminism, as I have discussed in earlier chapters. Therefore, I will presume that the object is moving toward a destination, and it is interacting strongly with that destination, which means that the destination already seems 'closer' than it is. Conversely, due to the exclusive focus of the moving object upon one state or object (i.e. the destination), the interaction strength with other objects must weaken. That weakening will reduce the contrast between the objects interacting weakly, which will translate into the increased *perceived distance* to the moving object. As this distance increases, the perceived *size* of that object decreases (that is now called length contraction), and the perceived time to reach the object increases (that is now called time dilation). Remember that the contrast reduces only in the direction of motion; length contraction and time dilation only occur in the direction of motion (not in all directions).

This idea can be analogously explained through the example of a human interaction. Suppose you are in a group of people, but you are focused on talking to one person, who is currently facing you. As you engage in intimate conversation with this person, others will feel that you have moved away from them—in the direction of motion, i.e. speech. You may still stand close to them, hold their hands, or occasionally look at them. Since you are now farther from them in one respect, therefore, you are *smaller* in that dimension. During this conversation, if others ask you a question, then you might respond, but only after a delay, since you are primarily attending to another conversation. This delay in response can make others think that if your responses have slowed, then, your clock must be running *slower*.

If this sounds a little silly and ridiculous, then that is indeed my intention. The ideas of length contraction and time dilation are perceived changes to the observed object when nothing has changed. But since everyone perceives in the same way, therefore, the effects can be verified empirically. The real question is: Why do such effects exist? The effect of length contraction is due to the perspectival nature of our perception. We think that those who are interacting with us are 'closer' to us, and those who are not interacting with others are 'farther' from us. Perspective tells us that things that are farther are also smaller, and things that are closer are also bigger. The perception of smaller size is the effect of perceptual perspective. The effect of time dilation arises because when we give greater attention to one fact, then we give lesser attention to other facts. The reduced attention to one set of facts will result in delays in the responses to requests. And if responses to requests are delayed, then we might claim that the clock has slowed.

Now, you might argue that these examples of human perception are misplaced because the effects of relativity are only seen at very high speeds. But what is speed? Relativity says it must be physical change in position. But it need not be. Speed can be the side effect of the changing strength and power of an interaction. If the interaction strength and power are changing rapidly, then the object is moving fast; if the interaction strength and power are changing slowly, then the object is moving slow. A faster speed means that the amount of information that we were exchanging is going down rapidly, and a slower speed means that the amount of information that we were exchanging is going down slowly. It is a fact of our perception that moving things catch our attention. Why? The reason is that there is a sudden drop in the level of information that we were receiving earlier, and that drop then prompts us to collect more information, which is an euphemism for saying that our attention is drawn toward the moving object. Similarly, if you only slightly shift your attention to one person, without neglecting the others, then the others will not consider you as distant (and hence smaller) as they would if you exclusively focus your attention to one person. Likewise, if you slowly reduce the interaction with one person during a conversation, then they will not consider you farther than if you suddenly cut them off. These effects are commonly perceived by everyone, and I'm only claiming that the underlying mechanisms for relativity and such ordinary human phenomena are the same. They are basically artifacts of our conscious experience.

Consciousness and Relativity

The essence of consciousness is *attention*. To be conscious of something is to be attending to or focused on something. However, this attention or focus also implies lack of attention or defocusing from other things. The more concentrated the focus on one thing, the less attentive we are to other things. As the attention to other things drops, the responsiveness to the demands, requests, calls, and invitations, also reduces. Thus, other people may send us 'requests' for information, but we may not respond to them, or our responses may be delayed. What happens when responses to our requests take more time? Naturally, we might think that the response is taking a longer time either because the person is far from us, and the time is being spend in transmitting and receiving the message. An alternative interpretation of this delay is that the responses are delayed due to a slower clock. Of course, in everyday life, we don't attribute delayed responses to either of these factors. We just know that the person is busy with other things, and therefore, not attending to our requests. But what happens when *people* are replaced with *things*? We stop thinking that they are not 'attending' to us. We start thinking that they must have moved far (or, equivalently, that their clocks are running slower). Relativity is thus the ordinary fact of defocusing transformed into a claim about sizes reducing and times elongating.

The mystery surrounding this idea would disappear if we thought of this problem in terms of consciousness and its impacts on perception. Namely, that when we attend to sight, then we don't simultaneously attend to sound. When we look at one thing closely, then we don't look at other things, simultaneously. Why? We cannot answer this question because our consciousness can only observe one thing at a time. It is completely different from classical physical objects which can interact with infinite objects in the universe simultaneously. Alternately, we can say that classical physics assumed that all objects interact with each other uniformly because its intuitions were grounded in billiard balls, and then found that these intuitions don't always work. Then, since relativity was superimposed upon classical physics, the formulation now looks less intuitive. Conversely, if our intuitions are grounded in conscious perception, and how observers interact with the world, then an alternative view about the world is generated. Now, we don't have to think that the lengths have shrunk, or the clocks have slowed down, but that these are merely consequences of change in *focus*. The latter intuition, I believe, is more instinctive and therefore we must reject the notions of physical objectivity and embrace the ideas of consciousness.

Physical relativity leads us to the rejection of the idea of an absolute space and time as an illusion, while the intuitions of consciousness lead us to accept an absolute space and time, and motion becomes an illusion. Therefore, in one sense, these two intuitions are equivalent reinterpretations of the same data. But in another sense, the intuitions about consciousness also lead us to completely different ideas about natural laws. Nature evolves not due to force but due to choices and the associated responsibilities.

For example, if we recall the conversation from the previous chapters, then each context requires a different focus—we may focus on some goals, but we may not necessarily neglect other responsibilities. Thus, the effects of conscious focus cannot be universal, as we can choose to focus and defocus at different times to different extents on different things. That changing focus would—if we see the world through the lens of physical relativity—entail that length contraction and time dilation cannot be universally true to the same extent. Indeed, if we believe in physical relativity, it will lead to the idea that the speed of light is not a universal *constant*, and that will in turn lead to the collapse of ideas that led to relativity in the first place.

On the other hand, if we think in terms of conscious experience, then the varying length contractions and time dilations would simply be the outcome of responsible or irresponsible choices, and those *results* would not be problematic, because each such result has a different *effect* and *consequence*, governed by a different kind of law. In short, collapse of physical relativity is not a problem, as it can pave the path to a better understanding of reality in which possibilities are selected by a choice, the choices are driven by goals, but the choices must also be contextually responsible, and responsible and irresponsible choices change and expand/contract our freedom.

In a limited sense, a consciousness-based viewpoint seems to merely reinterpret the observed facts, but this reinterpretation also anticipates when and how the observed facts will deviate, and how those inconsistent data sets can be reconciled with an alternative that works in both cases.

The Meaning of the Speed of Light

The constant speed of light has been debated since the 1905 paper by Einstein, owing to two reasons. First, there is no reliable way to synchronize clocks across space without assuming that space is *isotropic*—i.e. all directions in space behave the same; in the case of the speed of light, this means that the speed is the same in all directions. Second, there can be no reliable proof that space is isotropic, using linear speed measurements; since light's speed is linear, therefore, we cannot measure the speed unless we assume that space is isotropic, and there is no way to now that it is indeed so. In short, all measurements of speed of light rely on the premise that space is isotropic, but the fact is that the speed of light could be half in one direction and infinite in the other direction, and we would never know the difference. This is the continuation of the problem of homogeneity of space that we discussed in the last chapter, where we noted that there is absolutely no way to determine if space is homogeneous and isotropic and time is homogeneous. In the case of the speed of light, there are alternate theories at the present that produce identical predictions with the speed of light being half in one direction and infinite in the other directions. Numerous other possible speeds of light in different directions also produce the same empirical results. In that sense, what the real speed of light is remains an unanswered question. It could be any of the above values, and we cannot know for sure. We discussed in the last chapter that there are simply no empirical methods of verifying the properties of space and time from within space and time.

To add to this problem, we also know that the speed of light is not constant (even assuming space is homogeneous and isotropic) if light is not a *planar wave*. A plane wave is a sinusoidal wave, and light's speed depends on it being sinusoidal. If light is passed through different media, then its sinusoidal structure can be modified, and then it doesn't have a constant speed. In simple terms, the speed of light may not always be *measured* to be constant, if the amount of information carried in light is varied. The presence of non-sinusoidal patterns, for example, entails more information.

This fact assumes great importance if the cause of length contraction and time dilation is the changing focus of the observer, due to which variable amounts of information is transacted at different times. As some observations are prioritized over others, the delays in the other observations must increase, which will then entail variable measures of time dilation and length contraction. However, this also opens the door to saying that we need not measure the speed of light as *speed*, but as the amount of information being transacted. If the focus of consciousness changes, then varying amounts of information would be exchanged and that would give the appearance of a variable speed of light. But if we look at the problem from the perspective of an observer, then we can explain why this variation must be observed: the total amount of information exchanged is still constant, although that exchange is distributed over the transactions between multiple parties.

In short, the pairwise determinations of speeds between a sender and a receiver can vary, but the cumulative information received at each observer would still be constant (subject to the condition that they are exchanging information, which is not necessary since consciousness can withdraw from all interactions; one characteristic of choice is not to exercise choice).

Therefore, considering the above observations, we must rethink the nature of light, specifically, what does it mean to say that the *fastest* rate of information exchange is a constant? This conversation can be undertaken with two distinct ideas of information—*physical* and *semantic*. A fastest physical information rate simply means that there is a limit on the maximum *bitrate*; that is, we cannot exchange a greater number of symbols in a unit of time, regardless of the meaning that these symbols encode. A fastest semantic information rate, however, means that there is a limit on the maximum *idea rate*; that is, we cannot exchange a greater number of ideas in a unit of time, regardless of amount of energy being exchanged. The basic difference between the physical and semantic views is that the former focuses on the rate of symbol transfer and the latter upon the rate of meaning transfer.

We have discussed why a smallest number of symbols are needed for any idea to be exchanged. So, the problem of the fastest rate of symbol transfer must be complemented with the smallest number of symbols needed for the simplest of ideas. Let's call the maximum rate of symbol transfer R and the smallest amount of energy needed for an idea S. Now, R/S denotes the maximum rate for ideas. Likewise, if we fix the maximum rate of idea transfer as T, then T = R/S. Essentially, the semantic and physical notions of rates of symbol and meaning transfer are related by the notion of the smallest number of symbols needed for the simplest of ideas.

Therefore, both physical and semantic notions of the speed of light become subsidiary ideas to a more fundamental idea, namely that there is a simplest idea, which must be carried by one symbol, and the limit on the simplicity of ideas also defines the limit on the smallest number of physical entities. In short, we cannot talk about an idea being simple, if it cannot be exchanged as a unit. Similarly, we cannot talk about the smallest number of physical entities—which is one—if they do not encode meaningful information. Thus, we can talk about *fundamental* symbols that carry the simplest of ideas, and all complex ideas must be constructed from the simplest of these symbols. This symbol of the simplest idea defines our 'atomism'.

Once the smallest symbol of information is defined, then we can talk about the rate of information transfer. This rate depends on the source and destination of information. Physically, a system must be capable of transmitting and receiving symbols at a certain rate. Semantically, the system must be capable of sending and receiving meaning at a certain rate. However, the transmitted meanings are not *lost* by the sender, although they are *gained* by the receiver. (The simple intuition underlying this idea is that if you write your thoughts on paper, or utter them into words, you don't become ignorant of what you have written). When meaning is received, then it must be assimilated and organized into a body of knowledge. Each unit of information can alter the body, and the assimilation reorganizes the body.

Thus, the *maximum* rate of information exchange means that there are intelligent systems that can reorganize the body of knowledge at a fastest possible rate. Other less intelligent systems need more time to organize the body of knowledge. And knowledge cannot be transmitted faster than it can be assimilated. Therefore, there is a fastest rate of information exchange, but there is no slowest rate. Likewise, there are the simplest ideas that require the least number of symbols, but the ideas can get more complex.

Thus, the speed of light must be divided into two important ideas: (1) the idea of the smallest unit of information, and (2) the idea of the fastest rate of assimilating the smallest unit of information. Unless we define the smallest unit, we cannot define the fastest rate. Therefore, the smallest unit is a more fundamental idea, and the fastest rate rides upon that idea.

Once we decouple these two ideas, then we have a new way of combining them. The method of combination can be that atomic theory speaks about the smallest unit of information, while relativity speaks about the fastest rate of exchanging the smallest unit of information. The smallest unit of information is a property of space, because this unit must be encoded as a symbol, and constitutes a *location* in space. The fastest rate of information exchange is a property of time because the rate depends on how quickly we can assimilate information into a body of knowledge. This body also exists in space, and the fastest rate of information assimilation is the rate at which a collection of oscillators can be reorganized into a new structure.

We can illustrate this idea by an example. Suppose a student is learning a new subject from a teacher. He doesn't have the basic ideas required to grasp the subject, so the initial phase of learning generally goes slow because the new ideas must be absorbed, assimilated, and organized into a body of knowledge. The lower limit on the rate of learning is zero, i.e., the student may not learn the subject despite the teacher's attempts. Conversely, after the basic ideas are understood, then the rate of learning increases. However, there is an upper limit to the maximum rate of learning. This maximum rate of information acquisition represents the speed of light, but it requires the sender and receiver of information (e.g. a teacher and a student) to be conceptually similar. If there is great conceptual dissimilarity between the two, then the rate of information transfer declines proportionately to the dissimilarity. If we represent this dissimilarity by a semantic distance (in a semantic space), then far off entities in this space will exchange information slowly and nearby entities in the space will exchange information rapidly. As a result, the highest possible speed of information exchange will change. Now, the measurement of the constant speed of light is essentially a function of the similarity of the detectors between which the light is measured. If they are semantically similar (i.e. being used for the same purpose, which in this case, is measuring the speed of light), then the speed would be highest. This maximum speed is not a function of their distance of the speed of movement. It is simply a function of the fact that they are semantically *close*.

If space was infinitely divisible into infinitesimal points, then there would be no smallest idea or unit of information. Similarly, if time was infinitely divisible into infinitesimal points, then there would be no fastest rate of information exchange, because there would be no limit to the speed at which information can be reorganized into a new whole-part structure. Thus, the two properties of the smallest unit of information, and the fastest rate of information exchange, pertain to the *finite divisibility* of space and time. Thus, we can speak about the smallest 'atoms' or 'units' of space and time; the atoms of space encode the simplest or smallest meaning, while the atoms of time determine the fastest rate of reorganizing a structure.

Through this viewpoint, we can reduce the problem of the speed of light to the atomism of space and time—these two are not infinitely divisible, and their atomism represents informational properties of nature. The issue of atomism is more fully discussed in the next chapter, after we discuss the problems in quantum theory. But even the maximum limits on the speed of information exchange have important implications for atomism.

Takeaways from Non-Relativity

The discussion about whether space and time are relative or absolute, and the answers to these questions, leads to key ideas that will prove useful later in the book. We can recapitulate these problems and solutions:

- Inertial motion in classical physics requires that relative speeds be computed by adding and subtracting the speeds of moving objects, but this doesn't work with the constant observed speed of light. This discrepancy leads to the notion that moving objects undergo length contraction and time dilation, thus leading to the collapse of the classical objectivity that was associated with space and time.

- The failure of the objectivity regarding space and time, however, doesn't extend into the failure of the objectivity of other properties such as mass and temperature. And the laws of relativity are extended to accommodate the objectivity of physical properties along with the relativity of the measures of duration and distance.

- This reconciliation creates the Twin Paradox, where the objectivity of physical properties indicates that the twins must age the same, while the relativity of space and time indicates that one of the two twins must have a faster clock. This paradox is false because both clocks run at the same rate, and the twins do not age faster. The failure of the Twin Paradox shows the inherent problems of relativizing one set of properties while keeping another set of properties absolute, because there is no good way to separate them.

- Once the ideas of relativity are shown to be problematic, then we turn to explaining them in a way that preserves the facts, while resolving the contradictions. The cornerstone of this explanation is that *observed* distance and duration are byproducts of informational interactions between locations in the space of possibilities.

- We identify two properties—contrast and brightness, or strength and power—to explain how these interactions lead to the perception of length contraction and time dilation. We also see how these properties are initially understood as the facets of perspectival perception, and then simply as the attention of consciousness.

- Shifting attentions can change the length contraction and time dilation, and therefore the supposed constant speed of light. Hence, a new view of speed is proposed, that requires us to see space and time as being atomic; the units of space represent the simplest ideas, and units of time constitute the fastest rate of idea exchange. Thus, we reduce all properties to the atomism of space and time.

- The speed of light represents the maximum rate of transfer, which is maximized if the sender and receiver of information are very close. This is in effect like saying that if two people have the same ideological background, then they can understand each other very quickly. Those with different backgrounds, take more time.

- We continue to carry forward the notions of possibility and choice, the laws of responsibility, the application of responsibility in deciding the appropriate focus of consciousness, and how the space and time are 'objective'—i.e. different from the individual observer.

6

Is Time Discrete or Continuous?

Let's take a rocket ship
And blast off upon a trip,
A trip not among the stars, but in a stranger place!
A realm with the oddly thrilling name, Hilbert space!

Look how the basis vectors
Reach out to all the sectors!
This is the great achievement of the human race!
It's called Hilbert space!
—*Walter Fox Smith*

Problems in Quantum Theory

Classical physical theories of motion assume the continuity of space and time because such assumptions are required for calculus to be mathematically rigorous, and classical physics theories completely depend on calculus. Calculus, as it is used in physics, also requires first and second derivatives of a function to be continuous. Such derivatives are possible if the curves of motion are *smooth*—i.e. (1) there are no breaks in the curves of motion, and (2) the curves do not have discrete jumps. Thus, the general requirement in calculus is *smoothness* of curves, and the smoothness in turn assumes the continuity of space and time because without the space and time being continuous, the derivatives will also lose their usefulness. Classical physics theories also assume that matter is divisible into infinitesimal points. Thus, space, time, and matter are all comprised of *points*.

Each of these assumptions is challenged in quantum theory. Matter and energy must exist as discrete *quanta* rather than infinitesimal particles. The physical properties of the quanta are governed by Uncertainty Relations which stipulate that—(1) if the classical position is completely certain, then the classical momentum is completely uncertain, (2) if the classical time is completely certain, then the classical energy is completely uncertain, and (3) if the classical angle is completely certain, then the classical angular momentum is completely uncertain. While present atomic theory assumes that space and time are infinitesimally divisible, the uncertainty associated with such properties makes physicists question this divisibility in principle, although in practice the theory assumes a continuous space and time.

While discussing quantum theory, words must be used very carefully, because the same words have different meanings in classical and in quantum theories. One such set of words is 'position', 'time', and 'angle'. Quantum mechanics stipulates that atomic objects can have position, time, and angle *states*, but these states are discrete: all positions in space cannot be occupied; all instances in time don't exist; and all angles in space are not present. However, some position, time, and angle *states* exist, but the positions, angles, and time are "spread" and "discontinuous". Each location in space is scattered; each moment in time is a duration; and each angle in space is dispersed. Due to the extended nature of location, time, and angle, we can never reduce a quantum mechanical position state to a classical position—i.e. an infinitesimal point. Likewise, since a moment in time is extended, therefore, we can never reduce the quantum mechanical time to a classical moment. Finally, since each direction in space is dispersed, therefore, we cannot treat the quantum mechanical direction as a fixed angle. Since the positions, times, and angles are "discontinuous", we also cannot construct *smooth* trajectories connecting the positions, angles, and times. As a result, calculus cannot be used for quantum mechanical 'motion'. Quantum particles hence move from one state to another like a frog jumps from one puddle to another. The size of the puddle is never infinitesimal, and the puddles are discontinuous. This failure of classical concepts of motion, and the associated ideas of continuity and smoothness trouble physicists a lot, but truly understanding the quantum problem requires additional concepts.

The Notion of Orthogonality

The first such concept is that position, time, and angle states are *orthogonal*—a mathematical extrapolation of the ideas of spatial dimensions to algebraic functions. Just like the X, Y, and Z axes are mutually orthogonal, similarly, the position, time, and angle quantum states are also orthogonal. The true meaning of orthogonality becomes evident when quantum theory states that a quantum system exists as the *superposition*[2] of all orthogonal states, which means that only one such state of the system can be observed at any time. Other states exist too, but just like being on the X-axis means having no component of Y- and Z-axis, similarly, being in a quantum state means having no part of the other states. An observation reveals only one state at a time, which is like saying that the quantum particles are either on the X-, Y-, or Z-axes. Since each state has some energy, therefore, in an observation, the energy becomes visible, but it is otherwise hidden.

Second, due to the hidden nature of this energy unless an observation is performed, each orthogonal state is described as a *possibility*, with different *probabilities* which means that these possibilities are observed with different frequencies over time[3]. Since we can observe only one state at one time, therefore, the total *visible* energy of the system seems to change over time. We do not consider this a violation of energy conservation because the invisible energy exists in a hidden form. Due to the use of probabilities, the different quantum states are like the faces of a dice, and observation of a quantum system is like the rolling of a dice—only one face shows up. These energy states are 'potentials' which manifest one after another in time.

Third, unlike the dice, whose faces are separate, quantum states are said to be 'entangled'. What is entanglement? It is the existence of *individual* particles, that cannot be treated as *independent* particles[4]. What is individual and what is independent? Individuality means that we can measure a single particle at one time. Independence means that any particle can have any state or property. Quantum theory violates this intuition. Now, we have individual particles—which means that we can observe them as single particles. But these particles are not independent—which means that the states of these particles are collectively fixed. In a two-particle system, for example, knowing the state of one particle also fixes the state of the other particle. The property of 'entanglement' contradicts the independence of classical particles, so these quanta are particles, and yet not *classical* particles[5].

Fourth, if a particle is removed from a quantum system, then the states of all the remaining particles must change—compatible with the lower energy. Similarly, if a particle is added into a quantum system, then the states of all the particles collectively must be changed. After the addition or removal of particles, the new states would also be orthogonal. To maintain orthogonality, the energies of each of the other particles must be recomputed. In a sense, the addition or removal of some particle generally changes the energies of all the other particles in the system. In a sense, the total energy is *redistributed* in the system, and as the system becomes more complex, a small amount of energy addition or removal can produce drastic differences in energy states of the different particles. A classic example of this problem is the folding patterns of biomolecules; these molecules have biological functionality only when folded in a specific manner. A slight increase or decrease in the energy causes the folded proteins to become distended, lose their biological properties, or drastically change their functions. This is a quantum analogue of the classical 'Butterfly Effect' in which a small change in the input conditions—e.g. by the addition or removal of some energy—can produce drastic differences in the output. A quantum system thus can appear to be highly non-linear[6], although the theory is linear.

Since the quantum particles have probabilities, therefore, we can think of these particles collectively as a dice. Each face of the dice is a particle, but it is not independent. Therefore, all faces of the dice collectively constitute a single system. Since it is impossible to conceive of individuality without independence in classical terms, therefore, quantum theory seems hard.

We can however understand the quantum problem if we think of the quantum system and the particles in it as *words* and the quantum ensemble as *meaning*. The same meaning can be expressed using many words. A classic example of this problem is the existence of 'virtual particles'[7]. Then, even if a different set of words is used, no word is defined independently of the other words; rather, some words refer to other words which then refer to the referring words, which is why the words are 'entangled'. Finally, in any optimal language, only words that are unique in their meanings are employed, which refers to the 'orthogonality' of the particles. Indeed, if there are only two words in a system—e.g. hot and cold—then knowing one word tells you what the other word is. All quantum problems become easy if we think of a quantum ensemble as a meaning, the observation as the expression of this meaning into words.

The problem is that according to classical physics, matter has no meaning. Therefore, when words are described as meaningless particles, then all the properties such as entanglement, orthogonality, non-locality, etc. become very unintuitive. Even to say that a book has a certain probability for finding a word in it, is almost totally useless.

It is well-known that the world of concepts cannot be understood using classical logic, because it breaks all the logical principles. For instance, concepts are organized in a hierarchy and this hierarchy breaks the principle of identity; you can say that a "dog is a mammal" but you cannot say "a mammal is a dog". Likewise, something that is "not-dog" need not be a mammal; it can as well be a bird. This breaks the principle of mutual exclusion[8]. Finally, since a 'mammal' contains both 'dogs' and 'cats', therefore, in some sense it is both, which breaks the principle of non-contradiction. In effect, even if we can understand the quantum problems using ordinary concepts, we cannot use the classical logical and mathematical apparatus to model this reality. Quantum problem requires a logic and mathematics of concepts in which it is possible to add and subtract meanings to other meanings.

Thus, despite dozens of interpretations of quantum theory, the problem remains unsolved because every interpretation ultimately harks back to physical intuitions, classical logic, and mathematical formulations using number theory and calculus, all of which are inadequate and detrimental to the problem. Quantum theory needs a conceptual reality governed by a non-classical logic[9]. Without these, there can be no satisfactory solution.

Continuous and Discrete Times

Time appears in quantum theory in the two ways in which time is used in quantum *theory* and the *experiment*. The theory describes the *evolution* of the dice itself—i.e. how many faces it has. The experiment describes the *tossing* of the dice—i.e. which face of the dice turns up upon a measurement.

$$i\hbar \frac{\partial \psi\,(r,t)}{\partial t} = -\frac{\hbar^2}{2m}\nabla^2\psi\,(r,t) + V(r)\psi\,(r,t)$$

The evolution of the dice is governed by Schrodinger's Equation (given above), and it says that if a measurement is not performed, then the dice 'expands' over time. What is this expansion? It has no classical meaning, but if we take the view that the possibilities in the quantum state represent ideas, then the 'expansion' refers to the possibilities becoming more *different* over time. For example, when 'vehicles' were invented, they only included cars; but over time, cars diversified into trucks, vans, sedans, hatchbacks, scooters, motorcycles, etc. In the same way, Schrodinger's Equation says that even if we begin with an original meaning of 'vehicle', its *representation* will expand over time—i.e. it can be known in many diverse ways[10]. According to Schrodinger's Equation, there is no limit this expansion, but since the universe is finite, therefore, limitless expansion is practically impossible, hence, ultimately the quantum is absorbed by another system. For example, at some time, the idea of 'vehicle' may be subsumed under 'aircraft' as people leave the land for pedestrians and only use automation for flying.

The tossing of the dice is unpredictable in quantum theory—and it is not clear if the dice tosses itself, or it requires conscious intervention, or that there is some yet unknown mechanism that tosses the dice and then produces an outcome. There are some partial mitigations of this problem, that can sometimes *seem* to solve the problem. One such mitigation is called *decoherence* in which two systems—when brought into interaction with each other—can select states in each of the systems. The situation is comparable to two friends, who are undecided about what they want to do in their free time. One friend needs to choose between playing chess or going bowling, while the other friend needs to choose between going bowling or playing cricket. If these friends play with each other, then they select bowling, and the alternatives of cricket and chess are eliminated by their mutual company. The problem is: What brings the friends together? And which two friends must be brought together by which cause, accident, or serendipity?

Another promising alternative is Quantum Field Theory, where the probabilities of the alternatives change as an external force field is applied to a quantum system, and as the probability of some alternative becomes 1, the probabilities of other states become zero. There are variations of this solution, such as the Quantum Zeno Effect[11], but they are not satisfactory solutions either, because the basic question is still: What cause, accident, or serendipity causes the force field to be applied, when that force field must itself be treated quantum mechanically, i.e. in a superposed state?

Despite these difficulties in explaining how a superposed state becomes a singleton state, the fact is that singleton states are observed. And because these states are discrete, therefore, the time of our observation is discontinuous and *discrete*. Conversely, the time of Schrodinger's Equation, which determines the evolution of the quantum state independent of our observation is continuous and *smooth*. Thus, we are led to a fundamental paradox in understanding quantum theory, which is that we have two notions of time: (1) the objective time is continuous, and (2) the observed time is discrete.

If we rejected the continuous time, then we would be left with no predictions, because mathematical equations require continuity. Conversely, if we rejected discrete time, then we will produce a contradiction with observations. So, continuous time is necessary for the *theory* and discrete time for the *experiment*. Obviously, the goal of science is to explain experiences, so discreteness can never be rejected. Therefore, the theory must be rejected, but that rejection also involves rejecting mathematical equations. This is such a profound change, that it requires us to rethink everything.

A Solution to the Quantum Problem

To describe a solution to the quantum problem, I will distinguish between four levels of existence, which I will call—*possibility, capability, potentiality,* and *reality*. The possibility is a pure idea, and a reality is an experience. The pure idea transforms into an experience through two other steps— capability and potentiality. This transformation is depicted in Figure-26. Let's understand these four stages of existence one by and one, and then I shall try to connect them to the problem of quantum theory, and its solution.

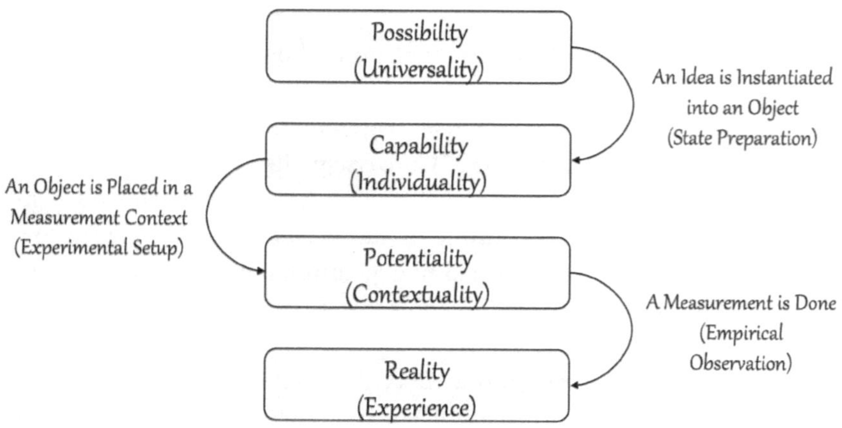

Figure-26 The Conversion of Possibility to Reality

The term *possibility* represents a pure idea; it exists eternally in the sense that all mathematical theorems, which may or may not have been discovered yet, are eternally real. We can also call this the Platonic world of pure ideas; although the Platonic world only comprises the *ideals*, in this case, even non-ideal ideas can be said to exist eternally. These ideas are not necessarily *true*. For example, there is an eternal thought that "the sky is purple". This thought is false, but a valid, meaningful proposition, and it exists eternally in that people in the past, present, and future can have the same thought. While having a false idea, they are merely accessing an eternal possibility. Similarly, people learning about true ideas are also accessing the eternal possibility, and as we have discussed earlier, our consciousness reaches into the domain of possibility during dreams, imaginations, creative phases, etc. to access the eternal ideas. Since these ideas can be accessed by other persons, therefore, they are not confined to their awareness; hence they are *objective*: objectivity is defined simply as the transpersonal nature of these ideas.

In this state of idea-like existence, the possibilities have no effect on the world; they are causally inert. To have an effect, the ideas must be *instantiated* into an individual embodiment or symbol of the idea. The instantiation requires us to combine the universality of the idea with the individuality of an instance. We can call this instance a *capability* because once the idea has instantiated into an individual thing, then it has the capability to create change. When we think of the Pythagoras theorem, then there is an instance of that idea in our awareness. Our awareness is therefore an *individual,* which combines with the idea, and creates an instance of that idea—i.e., our thoughts about the Pythagoras theorem are also instances of this idea. Similarly, when we transform the mathematical ideas into a book on mathematics, then the idea is embodied or individualized: there is an *instance* of that idea that others can perceive and use. However, to the extent that others may not read the book, or even know of its existence, the individual book is just a capability that can be known and used but may not be.

For the idea-instances to have an effect, these objects must be brought into contact with other objects, which are also idea-instances. We can call this contact or relation between such idea-instances *contextuality* because each context reveals only a part of the properties of this object, and more than one context may be needed to know the idea-instance fully. For example, if you are sitting inside a car, then you cannot see the outside, and vice versa. To know the car fully, you must be both inside and outside—alternately, which needs different kinds of contexts, relations, or measurements. The capability represents the *whole*, while the contextual potentiality denotes the *parts*—which are revealed in different contexts. If an object has been placed into a context, then one specific part or aspect of that object's properties has become a potentiality—there is an opportunity to know the object in a specific manner, but that opportunity may or may not be realized. For example, if you sit inside a car with your eyes closed, then the opportunity of knowing the car from the inside is not utilized. Likewise, if you purchase a book, but never read it, then the knowledge in the book is never gained. The book may have many pages, and the knowledge on each page becomes a potential when the page is turned. Thus, the preparation done prior to performing a measurement changes the observed world in the sense that *if* a measurement is performed, then the reality is braced for an answer.

As the reader flips through the pages, and goes through the chapters in the book, the potentiality becomes a reality—i.e. an experience, and the reader knows the mathematical theorems. If a person sitting in a car opens his eyes, the potentiality of a certain aspect of the car (i.e. the car's inside) becomes a reality. Here, I'm using the term 'reality' to not mean something that lies 'behind' our perception (many philosophers use 'reality' in this way) but to say that whatever was previously a possibility, capability, and potentiality, is now a reality—i.e., it has become someone's real experience.

Now, we can use these four tiers of existence to understand the quantum problem. A quantum system must be described at three levels of existence—capability, potentiality, and reality—because it is already an individual thing, and it has crossed the stage where an idea is converted into a thing. Due to this conversion, we must say that the quantum system is an instance of meaning. If we wanted, we could include the idea-to-thing conversion as well, and call it *state preparation*, which encodes meaning in a quantum system. Once the system has been prepared, it becomes a *capability* in the sense that it can be observed, but observation would require putting the system into a relation with a measurement system. When a prepared system is put into a measurement context, then it becomes a *potential* for observation. Finally, the potential for observation is also converted into a *real* observation.

The quantum system is complex; it has many aspects like a book has many chapters, the chapters have paragraphs, the paragraphs have sentences, and the sentences have words. All these parts of the book are organized in a hierarchy, and the book is therefore a hierarchically structured semantic system. The book can, in a loose sense, be called the 'superposition' of all the chapters, paragraphs, sentences, and words, in the sense that all these realities are objectively existing, although only one chapter, paragraph, sentence, and word is read at a time. Since we read the book step by step, therefore, we only observe one aspect of the book at one time. That doesn't mean that the quantum system is uncertain; it just means that we haven't selected some specific chapter, paragraph, sentence, or word.

Most mathematicians and physicists can easily accept the idea that there is a world of idea-like possibilities. Pursuant to quantum theory, almost everyone also agrees that the quantum world exists as a possibility. And everyone can agree that there is a reality or experience obtained during a measurement. Thus, possibility, capability, and reality are accepted to be relatively unproblematic ideas. The problem is that this reality 'braces' for a measurement when it contacts a certain type of measurement setup. A certain capability in the entire set of capabilities is relevant to the measurement setup, *if* a measurement is performed, and that aspect of the capability becomes 'prepared' or 'excited' or 'about to manifest' during a measurement. In a sense, the measured system is 'aware' that it is about to be measured in a specific way, quite like you might be mentally prepared to answer difficult questions as you enter an interview, or you brace for a 'good morning' as you meet coworkers while walking to your workplace, or you enter a state of heightened alertness as you hurry through a dark alleyway on a silent night.

In the capability stage, there is no expectation, bracing, or anticipation. All the capabilities are thus equally likely to occur, and none of the capabilities are more likely to occur. However, in the potentiality stage, all the capabilities are *not* equally likely, because some capabilities are more likely. Thus, a world of probabilities—in which some things are more likely—is produced from a world of capability because an object is placed in a specific context. We must therefore distinguish between an *objective* and a *contextual* reality. By 'objective' I mean that something which is perpetually a capability in an object. And by 'contextual' I mean a subset of that full capability which is now predisposed toward certain kinds of anticipated outcomes.

The essence of the quantum problem is that we are compelled to look at nature not merely as a reality, but also as a possibility, capability, and potentiality. These words may seem to mean the same thing, but they are different. Bundling them together is one of the sources of the confusion about the problem. Similarly, not recognizing that even the capability emerges from an idea-like possibility, and thus, what we observe as reality must be treated as a symbol of meaning, makes the problem of entanglement intractable. Since this meaning exists hierarchically as a whole-part relation, therefore, the space of possibilities, capabilities, and potentialities must also be described hierarchically. In this hierarchical system, each object is a dimension, which quantum theory calls 'orthogonality', and since there are literally infinite particles, therefore, there are literally infinite dimensions. And yet, all these infinite dimensions exist within three dimensions, because a dimension is simply a *type* or a *concept* in relation to a higher-level concept.

Finally, when a measurement is performed, in one sense, we are measuring the whole, because we are measuring a *property* of the whole. In another sense, this property is one of the many alternately visible properties, so it is not the whole. Thus, by knowing the part, we know the whole, and yet not know the whole. These two ideas about the measurement seem contradictory only in a classical sense when the whole is comprised of independent parts. They are not contradictory if we view the system as whole and part, such that each part is meaningful in relation to the whole. Thus, knowledge of a part contributes to the knowledge of the whole, and the whole defines the part's meanings, or what we even call knowing of the parts itself.

Two Interpretations of Probability

Accordingly, there are two interpretations of probabilities that we can adopt, which are equally true, and yet only a part of the fully story. For example, we can say that the cause of probabilities is the bracing of the system for an anticipated type of measurement, therefore, the probabilities can be attributed to the measured system. Alternately, we can say that the measured system is braced for some potentials more than others due to the context of measurement, therefore, the probabilities must be attributed to the apparatus that measures the system. Both alternatives are true, but they are only partially true, because they equally contribute to the outcome.

The objective interpretation applies the probabilities back to the measured system. Since the measured system is like a book, therefore, the probabilities pertain to the frequencies of the occurrence of sentences, words, and alphabets. Therefore, the parts of the book are not equally probable. Rather, based on the meaning, some parts of the book recur with certain objective frequencies, and these frequencies constitute probabilities.

The subjective interpretation applies the probabilities to the observer. In this interpretation, all the chapters, paragraphs, sentences, and words of the book are equally probable, but some readers may spend more time on some parts of a book if they are unfamiliar with some ideas. The probability indicates the relative amount of time spent on one part of the book vs. another. This probability is a property of the reader and not of the book.

Both interpretations are correct, but they are not *simultaneously* and always true. Some readers read the book cover to cover, strictly following the sequence of chapters, paragraphs, sentences, and words. Their probabilities would follow the strict sequence of words in the book, which means that the probabilities of the words must evolve—if the initial chapters dwell on different ideas than the later chapters. Other readers might jump from one part of the book to another, seeking answers to specific questions, and then returning to unread chapters after their curiosities have been satisfied. Their probabilities will evolve depending on the types of answers they seek. In the quantum experiment, similarly, the measuring system can be *passive*, and simply receive answers to a single question: "What is the quantum system?" But the system can also be called *active*, if it seeks answers to specific questions, and the order of events is decided by the question sequence.

Once we understand these different approaches to measurement, then we can also flip the roles of 'measured' and 'measuring' systems. Who says that a battery of detectors in the 2-slit experiment is not a measured system, while the source of blackbody radiation is not a measuring system? The notion of measured and measuring systems is an artificial creation of classical physics, where a stark distinction between an *object* and an *instrument* was made. In the real world, no such hard and fast distinctions can be made, because both systems must be described by the same theory—in this case atomic theory. Since the same theory applies to both systems, therefore, we must say that the roles of 'measuring' and 'measured' systems can also be flipped. The correct description of the world is that there are two systems, that are engaged in 'conversation'. Each system can ask a question and expect an answer. And it is possible to know someone as much from the questions they ask, as we can know them from the answer they provide.

Present quantum theory should be viewed as the 'average' of all these types of interactions. Thus, there is a contribution to probability that arises from the different frequencies of words. There is a contribution to probability that arises from the fact that some readers are more interested in some topics over others. There is a contribution to probability that arises from the fact that some readers are more adept at some topics over others. And if some reader was asked to read a book by a teacher, assigning them certain pages, there is a contribution to probability due to this assignment. I'm sure there are potentially numerous other sources of probabilities. In none of these cases are probabilities good explanations of the observation, because there are always underlying causes—such as the structure of the book, what the reader wants to read, what he was told to read, what he can understand, etc.—which can be used to explain the observed time spent on certain parts of the book. But each of these explanations is contextual, not universal. The universality, if at all there must be one, can only be the time averaged value over a certain averaged set of readers with different ideas and goals.

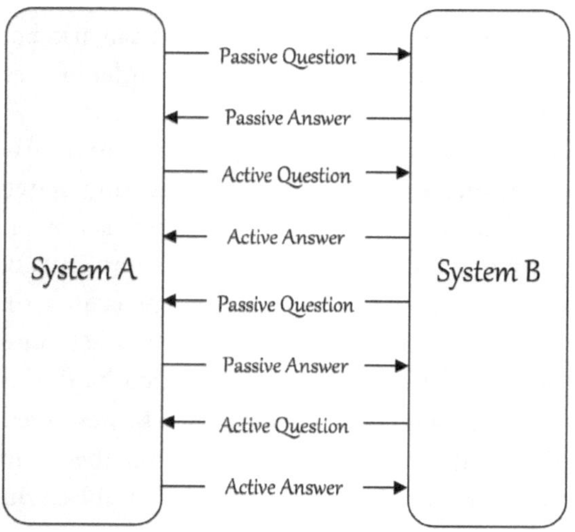

Figure-27 Many Possible Types of Semantic Interactions

I dwell on this point because numerous interpretations of quantum theory have tried to give a 'meaning' to the probability, *as if* that meaning would alleviate the problem. In my view, we don't need to interpret probability; we just need to reject it. If a system is physical, then all meanings are useless. But if the system is semantic, then all such interpretations are false. There is a real world with capabilities, some of which become more likely based on the context, but ultimately, what we observe is not due to likelihood. It is due to many types of interactions between the observer and the observed. And those interactions cannot be universalized, although that failure doesn't mean that there is no objective reality; it just means that the reality is not *objects*—i.e. things with no knowledge, no capability of knowledge, and no sense of awareness about the world. All the problems of quantum theory are outcomes of these classical physical dogmas, which are false, but because they have been so successful in creating technology, we are deeply attached to them. These dogmas have conditioned our thinking, and we are unable to think in a new way, because we only know how to think this way.

A Generalized Quantum Philosophy

To understand the true import of quantum theory, we must go back to the fundamental properties of a single quantum, which is said to constantly *expand* or *spread* (unless it is absorbed by another quantum). What is spreading? The possibilities in a quantum get *diversified* and the boundary of what we consider to be an 'ensemble' expands to include more and more possibilities. These possibilities, as we have spoken earlier, are eternal. Therefore, the expansion of the quantum is merely the expansion of a boundary to cover more of these possibilities over time. Unless, of course, a quantum encounters another similarly expanding boundary. Now, there are two possibilities: each of the expanding systems can absorb the other system or be absorbed. What is absorption? If we think in terms of a hierarchical whole-part relationship, then absorption of a quantum means that it becomes part of the other quantum, and the other quantum becomes a larger whole.

The property of entanglement in the quanta entails that once a quantum is absorbed, its boundaries are defined by the presence of the other possibilities within the whole. To the extent that the other quanta can also expand, the quanta that has newly become a part of a whole, can become a part of the part of the whole, or it can absorb the other quanta and become a partial whole, that contains the erstwhile parts of the whole. This process of reorganization of the parts—when a quantum is absorbed—is the essential import of entanglement: all the parts will be mutually orthogonal, so they cannot expand. But if they expand, then they can absorb other quanta and become larger quantum ensembles, with other quanta as their parts.

Thus, the basic process of the expansion of a quantum entails that each quantum tries to become an ensemble; but sometimes in the process of expansion, it becomes a part of a larger ensemble. Thus, the material world should be described as the 'struggle' for greater resources, 'competition' to become the ensemble instead of a part, and a 'conflict' for a superior position. The whole is the *dominant* quantum, and the part is the *subordinate* quantum. Every quantum is expanding, and therefore trying to absorb other quanta. By this process, it is also trying to attain a dominant position. Thus, matter only means competition, conflict, and a struggle of existence.

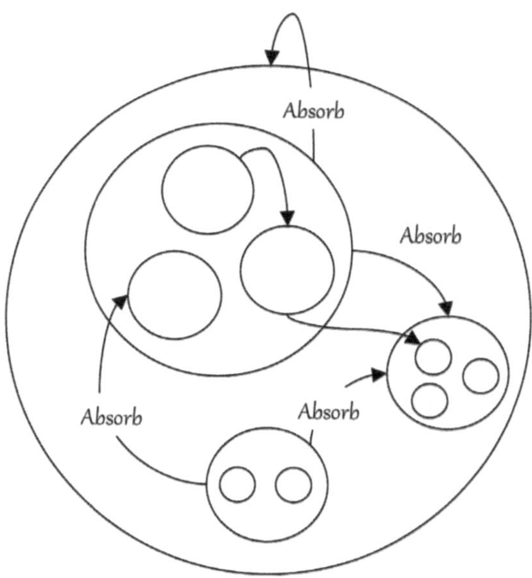

Figure-28 The Process of Expansion and Absorption

The *evolution* of matter is not truly the evolution of the possibilities; these possibilities are eternal. The evolution of the world is the expansion and contraction of boundaries in space; as a boundary expands, it absorbs other boundaries, and the parts within them, and becomes the *dominant* boundary, whereas the absorbed boundaries become the *subordinates*. The subordinates, however, don't necessary stay in the same state; they too must either absorb other ensembles, or be absorbed within other ensembles. The *competition* and *conflict* produce a dynamic in which what was earlier a whole can become a part of a larger whole, lose the parts to other parts, or even be fully absorbed within a part and become a part of the part. If this seems too complicated, then we can just think in terms of the whole-part hierarchy, in which the parts can become larger, the wholes can become smaller, the parts can become the wholes, and the wholes can become the parts.

The process of expansion and absorption presents a challenge for scientific thinking because the world now evolves due to competition and conflict, not due to consistency with the past states. As we know, mathematical models of causality assume that the future state is determined by the present state. But in this case, the future is determined by a *goal*—i.e. to absorb other quanta and ensembles. In current science, properties of particles are fixed, unless the particle splits or merges; but now, the properties of particles are redefined simply by their membership of a larger ensemble. The ensemble in current thinking, cannot become a part, such that the part becomes the ensemble; that would be logical contradiction in current thinking. Due to all these problems, we cannot form a mathematical theory of the quantum world in terms of current logic and/or mathematics. Logic is defined to be consistency with axioms, and it cannot deal with the idea of evolution through conflict and competition. The problem is therefore not merely physics; the problem extends into mathematics and ultimately logic.

We are also faced with certain kinds of indeterminism in this view. For example, when two quanta interact, which quanta absorbs the other quanta? This problem must remain completely random in current quantum theory, but it can be understood if we invoke the previous discussion about how responsible choices lead to consequences and effects that expand our freedom, whereas irresponsible choices contract the freedom. The conception of a natural law as the *responsibility* of choice is identical to the idea that one quantum must absorb the other: the absorbing quantum is 'expanding' while the absorbed quantum is 'contracting'. Unlike the present theory where we cannot predict which quantum will be absorbed, and which quantum will absorb, this problem becomes solvable through laws of responsibility.

But indeterminism never goes away completely, because these are the laws of *choice*, and not determinism. And yet choices are not random events; as we have discussed earlier, a choice is made through a process of rational minimization of costs and maximization of values. The responsibility of choice is that certain types of costs cannot be minimized, as they constitute our *duty*. For example, an employer must pay his employees a fair salary for their work; this is indeed a cost, and one can speak about 'rationality' that cuts out this cost. Likewise, protecting the environment necessarily increases the costs of production; this is indeed a cost, and one can talk about 'rationality' that eliminates these costs. Since rationality is prone to misuse, because it can minimize the costs that *should not be minimized*, therefore, (1) the choice of misuse never goes away, and (2) the consequences and effects of such choices must follow naturally (i.e. not via man-made laws).

Thus, the idea of universal possibilities is not contradictory to the ideas of the boundaries of individuals, the choices of goals, and their effects and consequences, and that these laws in turn dictate which ensemble becomes bigger, and which ensemble becomes smaller, or which quantum becomes an ensemble, and which ensemble becomes the quantum. The real dynamic of the quantum world cannot be separated from the ideas of competition, conflict, choice, responsibility, and consequences. It is understandable if this seems overwhelming. But unless we know the solution to a problem, we don't necessarily know how big that problem is. This is indeed true for the quantum problem. But in another sense, it is only a small part of a broader gamut of problems about matter, causality, conscious observation, etc.

The Discreteness of Time

We can now return to our original question: Is time discrete or continuous? In present atomic theory, the event observations are discrete, and Schrodinger's Equation, that calculates probabilities, is continuous. If these probabilities are rejected, then the continuity would disappear from quantum theory. We would ask: Does the underlying reality of possibility *change* in discrete steps or continuously? And the answer is that all change occurs through informational interactions, which are always discrete. Therefore, the results of these interactions are discrete, and since a finite time is required to complete a discrete transaction, therefore, time is discrete.

This is a profound change in science, and it means the overturning of calculus, and thereby of all mathematical laws that rely on calculus. We have already seen that the quantum dynamics requires goals in the future, because conflict and competition cannot exist without a goal. Likewise, the winners and losers of a competition cannot be decided without considering the preparation of the competitors in the past. Each moment in time is discrete because what we call *change* is the *flipping* of the whole-part relations, and something must either be a whole or a part, but not something in between. In short, we cannot say that a whole absorbs a part through infinite infinitesimal steps. This is already evidenced in quantum theory: a quantum is either fully absorbed, or not absorbed at all; the quantum is not gradually and incrementally absorbed. Thus, the events of absorption, emission, and reabsorption constitute discrete changes to the whole part structure, and not only do discrete changes take time, but they are also discontinuous.

Thus, the causation of an event—which arise due to choices—is discrete. And the effects of this causation—i.e. change—is also discrete. Whether we measure time by the cause or the effect, the conclusion is the same: time cannot be considered continuous; it must be discontinuous.

Takeaways from Discreteness

The discussion about whether space and time are discrete or continuous, and the answers to these questions leads us to some key ideas about matter and its dynamics. We can summarize these ideas as follows:

- Quantum theory has an innate contradiction due to the two ways in which it describes space and time. In Schrodinger's Equation, space and time are continuous, but in the empirical observations, the space and time are discontinuous. This contradiction is resolved by the use probabilities, which lead to problems of causality.
- The fact that probabilities work, however, indicates a new conception of reality, which is that our observations of the world must be distinguished from the world as it exists prior to observation. This is addressed by the distinction between 'potentiality' and 'reality'.

- Potentiality and reality, however, don't explain why quantum particles are entangled and yet distinct: they are *individual* entities, and yet, they are not *independent* entities. The resolution of this problem requires semantics with its notions of wholes and parts.

- When we combine all these ideas, we get the notion of four tiers of reality—possibility, capability, potentiality, and reality. The first is universal, the second is individual, and the third is contextual. The fourth constitutes experience, which is relative to an observer, and hence the principles of relativity apply to the observations.

- Probabilities, however, constitute an unsatisfactory aspect of quantum theory. They can be replaced in a semantic conception of reality where the probabilities turn out to be averaged properties of many real but diverse scenarios involving the ability in the quantum systems to be 'aware' of the presence of other systems, represent meanings, and the capacity to transact informational messages.

- Once the problems of probability and entanglement are addressed by the use of semantics, then we can construe a picture of the quantum reality, and this is facilitated by the idea that an isolate quantum always expands until it absorbs another quantum or is absorbed by it. The process of expansion thus leads to the idea of competition and conflict between which quantum will absorb other quanta.

- At this stage, the outcome of the conflict seems random, but this randomness can be overcome if we recall the previous ideas about the possibilities expanding and contracting as the effect and consequence of previous responsible and irresponsible choices. Therefore, the outcome of the conflict is also logically predictable.

- When we put all these conclusions together—i.e. that absorption is discrete, the quanta as possibilities are discrete, and the evolution of the wavefunction[12] occurs due to the absorption or emission of discrete possibilities into whole-part relations—then, we arrive at the conclusion that space and time must also be discrete.

- Discreteness simply means that the cause of change is an objective goal that determines choices, and the effect of this choice is a discrete state change due to flipping of the whole-part relations. Therefore, time is discrete both as a cause and as an effect.

7

Is Time Reversible or Irreversible?

The Second Law of Thermodynamics

In earlier chapters, we discussed the use of shortest-path descriptions of nature, and I described how classical mechanical theories comply to such doctrines of optimality. A shortest path between two points also creates a *reversible* process, in the sense that the steps by which one went from state A to B, is also the path from which one can go from B to A. Since classical physical theories comply to the shortest-path principle, therefore, these theories are also *reversible*. The determinism of Newton's mechanics fails in thermodynamics, and the reversible trajectories of individual particles create a collectively irreversible system trajectory. The irreversibility of thermodynamics arises because both least-action and shortest-path principles do not apply[1]. The total energy is still conserved, but a distinction between *useful* and *useless* energy must now be drawn—the *useful* energy is called "work".

Figure-29 illustrates this idea through pressure-volume (called P-V diagrams) and temperature-entropy (called T-S diagrams) cycles. An irreversible process constitutes a *cycle* in which the reverse path (3 -> 4 -> 1) is always longer than the forward path (1 -> 2 -> 3). In a purely reversible process, these two paths would be of equal lengths, which means that process could be represented by a square or a rectangle. When the forward and the reverse paths are not equal, then the shortest path principle is broken, and thermodynamics describes this problem as the continuous increase in *entropy*[2].

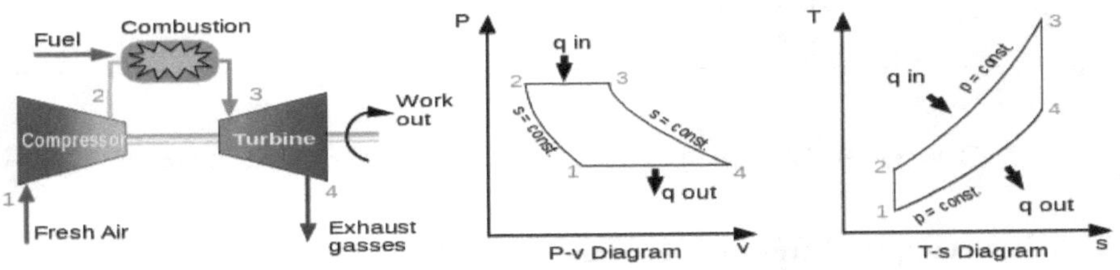

Figure-29 Hysteresis in Thermodynamic Cycles

We have already noted the failure of deterministic dynamics while describing particle collections: the total energy in a system can be redistributed in innumerable ways, consistent with the laws of classical mechanics. However, one assumption latent in classical mechanics is that only *one* of these innumerable possibilities is real. The Statistical Mechanics description of heat breaks this principle. It says that to explain the irreversibility of heat, we must assume that *all* the states can be real, and the system has a cross-state dynamics in addition to the evolution of trajectories within a state.

As we know, in classical mechanics, each particle has 3 position and 3 momentum states. Thus, a single particle can be described in a 6-dimensional space. When N such particles are placed in space, then 6N-dimensional space is required to describe *one* state. We can simplify this space as the two dimensions of 3N positions and 3N momenta. One state of the system is one point in this space. Then, as the system can be in literally infinite such states, therefore, we can present all the system states in the space of 3N positions and 3N momenta. These states have a *distribution* in the sense that in some places they are closely packed and in other places they are loosely packed. Now, we can suppose that the system has some dynamics in this 6N dimensional space—i.e. it moves from one system state to another. The succession of these models is depicted in Figure-30.

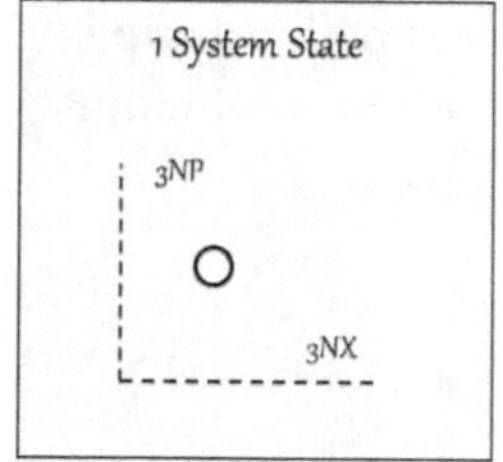

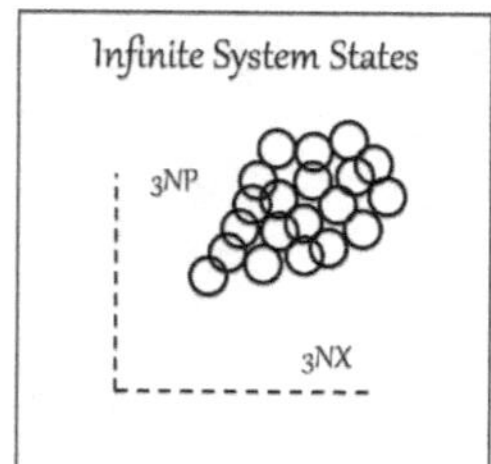

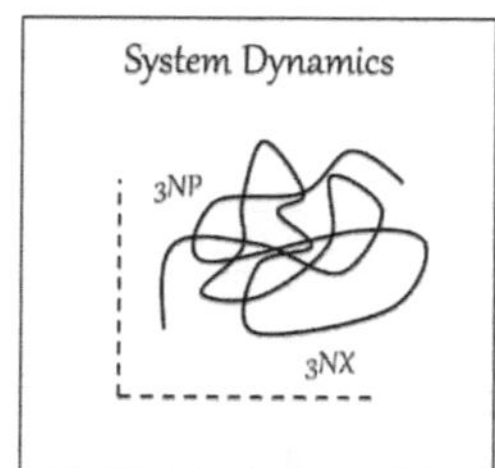

Figure-30 System Dynamics in a 6N Dimensional Space

For example, there is a state of the system in which all the particles are moving in a single direction, and another state of the system in which the particles are moving randomly. When an observation is performed, one of these states will be observed, and the state in which the particles are moving in a single direction will able to deliver more energy than the state in which the particles are moving randomly. According to classical statistics, each such state is equally likely, but there are many more disordered states (e.g. where all the particles are moving randomly) than there are ordered states (e.g. where they are moving in a single direction). Hence, if a measurement is performed, it is more likely to find that the system is in a disordered state than it is to find the system in an ordered state. As a result, due to the cross-state dynamics, and the disordered states being more likely, the *available* energy in the system goes down, even if the total energy is constant.

Classical and Quantum Statistics

Here, a distinction can be drawn between quantum and classical statistics. Think of some playdoh. In a quantum statistical measurement, a small piece of the playdoh is pulled out from a bigger chunk of playdoh and observed. This observation pertains to a *part* of the whole, and the probability pertains to which parts of the playdoh are more likely to be observed. In a classical statistical measurement, the *whole* playdoh is observed, but it is constantly changing its shape. Some of these shapes look like trucks and cars and are more useful. Other shapes are amorphous, and don't resemble anything useful. If we consider all the possible shapes of playdoh, then the useless forms are much more numerous than the useful forms. And in such cases, the work performed by the system is significantly reduced. In short, in quantum statistics, a dice is thrown, and one of the faces turns up; quantum probabilities pertain to which face turns up. In classical statistics, the dice changes its shape, and if the dice is thrown, we measure the whole dice; however, the disordered shapes are more likely than the ordered shapes, and the probabilities pertain to which kinds of shapes are more likely.

Classical statistics has a counterpart in a quantum system too, where the quantum *wavefunction* can be divided into many different *bases*. We can think of these bases as different dice shapes: the faces can be more and less, and each of the faces can have different shapes and sizes. The total energy is still constant, but the energy is *distributed* in the system in different ways. However, in a quantum system, the measurement procedure is fixed by the experimental setup. For example, if a 2-slit experiment is performed, one out of the innumerable shapes of the dice is chosen, and the subsequent measurements pertain to a specific face of the dice turning up, not to changing dice shapes. If classical statistics is also applied to a quantum system, then we would say that upon each observation, the number of slits in the system is also evolving, and as a result, the dice is changing shape, and the observations of faces turning up also pertain to a shape-shifting dice.

Quantum and classical statistical systems are therefore not mutually exclusive, although in a quantum experiment the classical statistical properties are hidden by an experimental setup, and in a classical measurement, the quantum statistical properties are hidden by the experimental setup. In effect, we are talking about two kinds of experiments—one those reveal the sides of a dice, and the other that show the changing shape of a dice.

The Violation of the Shortest Path

Now, we can say that if the dice was constantly changing shapes, then the *path* it took from shape A to B, may not be the path it takes from B to A.

For example, in state A, the dice could be an even six-faced cube. It might then change into a 7-face, 8-face, and eventually a 10-face object. Let's call this state B. When it traces the path back to the six-faced even state, it might not necessarily retrace its steps in reverse. It might also from a 10-face object to 8-face, 7-face, 9-face, and 6-face objects. When the reverse trajectory doesn't retrace the forward trajectory steps, then, more energy is spent in transforming an 10-face dice to 8-face, 7-face, 9-face, and 6-face dices, and the shortest path principle is broken—the forward and reverse paths are not identical, and the paths in both direction may not be shortest path[3].

However, this doesn't say that the reverse path will always be *longer* than the forward path. It is also possible that the forward path could be longer. This is when a more precise statement about natural processes is made: The reverse processes will always be *longer* or *equal* to the forward processes. This precise statement is called the "Second Law of Thermodynamics":

The total entropy of an isolated system can never decrease over time and is constant if and only if all processes are reversible.

Entropy represents the total number of possible states of a system, or the total number of shapes that a dice can take, given the same amount of playdoh. How can the total number of states increase? The answer is that it is possible if chunks of playdoh are combined, then the total number of shapes you can make out of a greater amount of playdoh is also greater. The standard example of such entropy increase is the mixing of gases. Suppose that there are two chambers containing N and M particles of gas. Then a 'door' between the chambers is opened and the gases are mixed. The result of this mixing is that the total number of possible states of the system increases significantly, because there are now N + M particles of gas.

But this conclusion is highly unintuitive in classical mechanics because who is to say that there are N and M particles of gas? From a classical mechanical perspective, the same energy could also be divided into a greater and lesser number of particles. As we have discussed earlier, when particles collide, they can coalesce or split. Similarly, we can make as many shapes with a small amount of playdoh as with a large amount of playdoh *if* the playdoh is divisible into point particles. There are as many points in one inch of length as there are in two inches. Therefore, there is no reason to suppose that there are N and M particles before they are mixed, and N + M particles after they are mixed. This assumption is valid *only* with quantum mechanics, or a limit to the divisibility of matter, because then, an inch has half the number of points as two inches. Therefore, classical statistical mechanics would fail unless we add premises of quantum mechanics. This goes back to my point that these two types of statistics can be combined.

Now, when we answer the question of how states increase, then we can also answer why they don't decrease. The reason is that matter is not infinitely divisible, so if we mix things, then the total number of possible states increases, and the trajectories are likelier to pass through more states, and that make them more likely 'longer'. When a system goes from state A to B through X steps, and from B to A in Y steps, and Y > X, then, the reverse path is longer. This increase entails that the *shortest-path* principle is broken, because the system meanders from one state to another through a longer set of steps than before, and this 'disorder' is due to higher 'entropy'.

The principle of least action requires the most optimal path to a destination, but such a path may not necessarily be chosen. If optimal paths are chosen, then the path is also the shortest, and the system is reversible. But if suboptimal paths are chosen, then the path is not the shortest, and the system is irreversible. As the number of states increases, the chances of a longer path increase, and the system becomes irreversible. The principle of least action *should* never be violated, but it *can* be violated. If it is violated, then forward and reverse paths are not equal, and an irreversible process is created. However, to the extent that it can be violated, it is now claimed that time has an 'arrow' or a direction, which is the direction of entropy increase. In short, as nature goes through cycles, each such cycle creates disorder, and the universe is therefore headed toward a 'heat death' in which all order will be destroyed, and the universe would become incapable of any work.

Maxwell's Demon

Much of the modern conversation around time now hinges upon the idea of time as something that emerges out of the increase in disorder. The conceptual basis of thermodynamics is that things naturally *mix*, but they don't naturally *unmix*. James Maxwell, in fact, conceived of a 'demon' who would selectively open a 'door' between two chambers containing mixed gases, allowing the red gas molecules to move into chamber A and the blue gas molecules to move into chamber B, such that the separation of gases into two chambers would create order instead of disorder. The selective opening and closing of the doors between chambers would require: (1) a cognitive capacity to understand which molecules are red or blue, and (2) the ability to close and open the door without any additional energy.

Such a 'demon' cannot exist in classical physics, because both the cognition of molecules, and the opening and shutting of the doors required the demon's cognitive and conative system be fed with energy, which would increase the entropy, even in the process of trying to decrease the entropy. The net result of these two processes will still be an increase in the disorder. Thus, Maxwell concluded that no demon could reduce entropy, and entropy increase is an inviolable law of nature. The continuous increase in entropy can now be thought of as the direction of time from order to disorder.

Problems in Maxwell's conclusion appear if we examine a living cell. It is well-known, for example, that the cells in living systems selectively ingest food and excrete waste. The molecules in the cell wall have recognition capacities by which they know the difference between food and waste. And the opening and closing of the doors (in this case the cell wall) is simply the reconfiguration of a molecule's folding pattern which creates a way for the food to be ingested and the waste to be excreted through the cell wall. This selective ingestion and excretion of food is just like Maxwell's demon which opens the door selectively for red or blue molecules, and hence it violates the second law of thermodynamics, or the principle of entropy increase.

What is going on? Why does the second law fail? This goes to the peculiarities of quantum reality. According to quantum theory, the total energy in a system can be divided into eigenfunction states in many ways. Each such way of representing the total energy is called an *eigenfunction basis*. In the slit experiment, these bases are selected by using a different number of slits. Thus, for instance, if 2-slits are used in the slit experiment, then the energy is distributed differently than if 3-slits are used. No energy transfer is required either to or from the slits to reconfigure the eigenfunction basis.

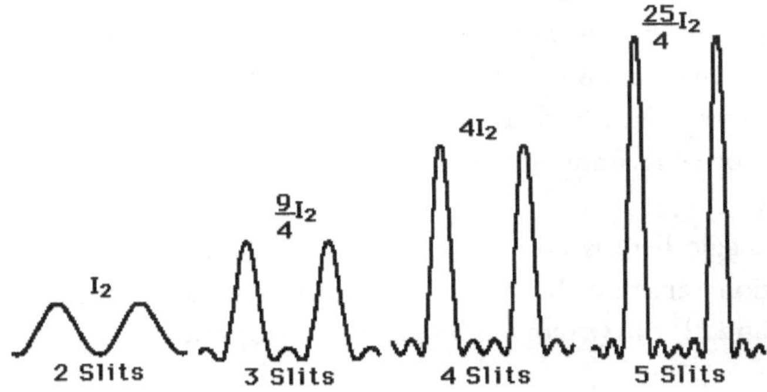

Figure-31 Change of Quantum Basis by Slit Selection

As we have discussed, this is like the dice changing its form, such that the same amount of playdoh can be reconfigured to create dices that have fewer but larger faces, as opposed to dice that have more but smaller faces. According to the principle of entropy increase, the dice can only increase its faces—i.e. have a greater number of possible states—but this principle is not upheld in quantum theory. Rather, it is possible to change the eigenfunction basis, and thereby the reduce the total number of states. To change the eigenfunction states, there must always be something that reconfigures quantum system—e.g., in the slit experiment, it is the number of slits.

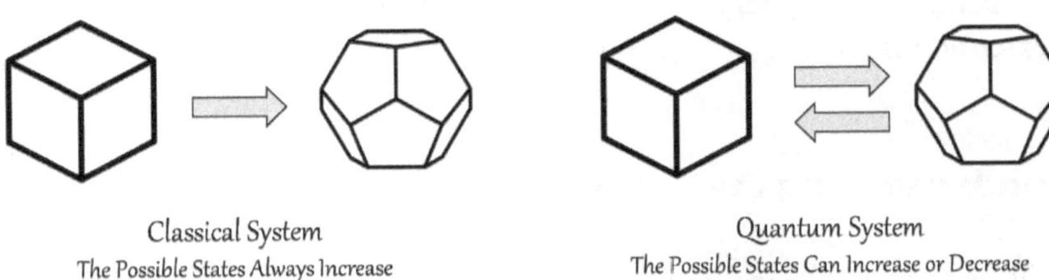

Figure-32 Quantum Theory Violates Entropy Increase

In the case a biological molecule, we can surmise that it is the presence of 'food' or 'waste'. If the mere presence of 'food' or 'waste' can reconfigure the cell wall, then it is the perfect analogue of Maxwell's Demon which opens and closes the door just based on it cognitive and conative abilities. In essence, there is a material system—food or waste, and a cell wall—but it changes its behavior *without* exchanging any energy, and if no energy is being exchanged, but order is being created, then entropy increase is not a universal law. Such increases would be facts about those systems that cannot be reconfigured just based on recognition and can be called 'classical'.

The magic here is created in two ways: (1) a quantum system has recognition capacities by which it can detect if something is 'food' or 'waste', and (2) the system has bootstrapping capacities to change its own structure.

Thermodynamics is the description of engines and turbines which don't have recognition and bootstrapping capacities. Thus, an operator must push a button to start the engine, and the engine must be fed some energy for it to change its state. Both these actions require external intervention, and thus, an observer can control the machine. However, in the process of controlling the system, an irreversible process can also be created, and if we think that all of nature is like an engine or a turbine, then entropy increase must be a universal law. But this isn't necessarily true.

We also cannot disregard the fact that the Earth has supported life for thousands of years because nature automatically recycles matter and energy, and whatever was previously mixed becomes unmixed. Nature's processes can thus create order. However, human inventions such as engines and turbines, are irreversible. The second law of thermodynamics is a frivolous extension of classical mechanics to all nature, and this extension is false.

Understanding the Intention

Once we reject the truth of the thermodynamic arrow of time, we are left with a new problem: We have no other explanation of the time direction in physical theories because all such physical theories seem reversible.

But we need not despair. There is a time direction in every cell because the cell ingests food and excretes waste, and the arrow of time depends on understanding the process by which a cell wall molecule reconfigures itself. How, for instance, does the cell wall know that something is 'food' vs. 'waste'? How does our immune system know that some bacteria are 'useful' rather than 'harmful'? After all, the immune system doesn't ingest the harmful bacteria, and allow it to produce damage before it determines that *this* specific bacterium is harmful; it knows that the bacterium is harmful before allowing the damage to occur. But how can recognition occur?

To understand this issue, we can look at a famous problem in computational theory, called Turing's Halting Problem. The Halting Problem says that a computer cannot determine what the program will do *before* it does it. One outcome of the problem is that the computer cannot determine if a program is harmful, *before* it causes the harm. As a result, computers are prone to viruses, because the operating system of the computer (also a program) cannot decide which software is malicious. After all, malice is an *intent*. To know if a program is malicious, we must know the intent of the program, but physical processes—such as those used computers—are incapable of deciding that. It is a recognition of the fact that current computers do not use quantum capacities to the fullest. As we noted above, a quantum system can reconfigure its eigenfunction basis based on the number of slits, which means that the quantum system 'braces' for a measurement. A truly quantum computer would also be able to recognize the *intent* in the other systems and be able to reconfigure itself in this way. As a result, a truly quantum system would not be prone to computer security problems seen today.

To explain the abilities in a quantum system to recognize other systems and brace itself for an eventuality, *before* that eventuality occurs, we must say that matter comprises intention, which is not difficult if we accept that a quantum system carries semantic information. This intent would however be different from the cognitive meaning. For instance, you can utter the same sentence in a neutral or sarcastic tone, and in the former case, it means that you agree with the cognitive meaning and in the latter case it means that you disagree with that meaning. Thus, the meaning of a sentence is not fixed by the structural sequence of the words, or the meaning of the individual words. It also requires an intentional component. That intent is not *outside* the sentence; it exists in the sound of the sentence, as the *tone* in which it is spoken. In the case of written text, where the tone is not available, the intent is understood only by use of exaggerated expressions. For instance, if someone asks you: "Do you think this is true?" then in spoken language, you might say "Of course!" sarcastically. But in written text, you must say: "Of course, I have to agree with everything you say!" where the exaggerated and vociferous expression of agreement factually signals a disagreement.

The point is this: We habitually decode intention from the written text, or from the spoken words. We don't take everything that is said literally. There is of course a literal meaning. But it is also nuanced by the contextual differences, and by the intention behind those words, and these three kinds of meanings must always be considered before we know the meaning.

The meaning of the sentence "I do" for instance depends on what was said previously, and to which you are now agreeing; therefore, if you think that the sentence indicates acceptance, then you could be wrong, or you may not have the fully meaning. Thus, for example, when a priest says: "Do you take so-and-so to be your lawfully wedded husband?" and you say, "I do", the meaning is that you are accepting marriage. But when a judge asks you, "Do you swear to speak the truth, all the truth, and nothing but the truth?" and you say, "I do", then the meaning is that you are agreeing to speak the truth. The meaning of the sentence "I do" therefore depends on what was said previously in the context, although it also has a universal meaning of acceptance. Similarly, if you said, "I do" sarcastically, mockingly, or jokingly, then by the tone, you would also be indicating the opposite: "I don't".

Thus, we must draw two crucial conclusions. First, that there are three kinds of meanings, which must be combined. Second, these three kinds of meanings can either enhance, diminish, or contradict the other meanings. Therefore, while combination is necessary, it doesn't always produce consistency. As a result, we also must *prioritize* the meanings, and that prioritization depends on some universal principles (e.g. that one must always tell the truth in front of a judge), some contextual principles (e.g. that certain things generally mean something in a specific context), and some individual principles (e.g. that this person generally means that if he uses some specific words). The problem is meaning is incredibly complex, but the point is that this complexity doesn't deter us from understanding universal, contextual, and individual meanings. We employ cognitive, relational, and emotional content to decipher the meaning. And we try to resolve the contradictions arising in between these three kinds of meanings in the best possible way. This doesn't entail that we will always know the truth. But it does entail that we will know much more than if we never tried to know the meaning.

Thus, our immune system protects us from numerous types of bacteria and viruses. But sometimes our immune system also produces an allergic reaction to useful things like nuts or dairy. And many times, our immune system fails to recognize the constantly evolving bacteria and viruses. All these cases can be understood if (1) we recognize the objective reality of intent—e.g. that the immune system anticipates what something is going to do before it does it, (2) it can sometimes be wrong in that anticipation, thus resulting in an allergic reaction, and (3) sometimes it fails to anticipate, thereby resulting in infections and diseases due to immune failures. Just because the immune system can sometimes judge something useful as something harmful, or judge something harmful as something useful, doesn't mean that these judgments are not protecting us in innumerable other cases, or that the cognition of an intentional reality itself doesn't exist. One of the main reasons that modern medicine only tries to fight diseases, rather than improve the immunity to diseases, is because the physical and chemical foundation of modern medicine is devoid of grasping an intentional reality. Since the dawn of science, it has been assumed that matter has no intention; that such things exist only in our minds, and even then, these intentions are merely the epiphenomena of chemical reactions, which are physical. Such theories, of course, will prove to be ineffective to (1) explain how the biological systems work, and (2) fail to identify ways to improve them.

Once we understand and accept the existence of an intentional reality, then we can say that matter itself has a purpose. The molecules in the cell wall have a purpose to 'defend' the cell. The white blood cells in the body have a purpose to 'fight' harmful bacteria and viruses. The DNA molecule also has a purpose to help the organism 'survive', because it selectively generates different kinds of proteins based on the current bodily condition, even though it can potentially generate every kind of protein always (a classic example of such generation is the synthesis of enzymes that digest food only when we eat, or those proteins that then generate stomach acid). Such varied responses to environmental conditions, which also vary in different individuals who might have the same genes, belie the premise that matter has no recognition, or that it doesn't operate based on intentions, or that the presence of a chemical always produces a predetermined outcome.

Linear vs. Cyclical Arrows of Time

The presence of intention produces a *directed behavior* and is the first cause of the time direction. For example, the cell wall is far more likely to ingest food and excrete waste than the other way around, because the molecules in the cell wall have an intention, and the capacity to recognize intention in other molecules. A white blood cell is far more likely to fight a harmful bacterium than a useful bacterium, because the cell has an intention to protect the organism, and the capacity to recognize intention in other cells. While a cell wall molecule can equally well ingest and excrete anything, it overwhelmingly acts in one direction vs. the other—due to the intention. Commonsensically we understand that *we* have a direction in life *if* we have a constructive goal. Thus, people with destructive goals are said to be going 'backward' in time, or that they are 'regressive' instead of 'progressive'. The arrow of time is the sense of progression, and it exists because there are intentions. We might have harmful intentions, but those are intentions too, and we act in a certain way because of those intentions. Thus, the first basic sense in which the time's arrow arises is due to presence of intents.

This intention, however, will not always work, unless these intents are also connected to each other through an *ecosystem* of interactions. For instance, for the body to continuously exhibit purposeful direction, 'waste' from one organism must become 'food' for another, and matter must be recycled to perpetuate the consumption of food and excretion of waste. This is indeed the observed fact in ecosystems where organisms with opposite types of appetites are interrelated. Through such relation, a cycle of food is created, and the cycle perpetuates since each spoke in that wheel moves forward for two reasons: (1) each spoke has a capacity to recognize food from waste, and fulfill its intent, and (2) all the spokes together are woven into a wheel, such that the movement of some spokes naturally pushes others.

When we recognize the intents in individual things, then we get a linear sense of time. However, when we recognize the fact that these linear intents cannot be sustained automatically over time (again, since the universe is finite), and their sustenance requires a cyclical ecosystem circulation, then the real nature of time is understood to be cyclical rather than linear. We can now say that time is not moving linearly forward from order to disorder. In fact, since 'waste' for one organism is 'food' for another, therefore, terms such as 'order' and 'disorder' have no absolute meaning. Each organism has their own definition of order and disorder, food and waste. Objectively speaking, nature simply has *forms*, and some forms are mutually compatible while others are not. Thus, a form considers some other form compatible, and hence 'food', but that same form is 'waste' for the other form. For example, fruits and grains are 'waste' for the trees and plants, whereas manure and carbon dioxide are 'waste' for humans. The correct description of nature is not that it moves from order to disorder or vice versa, but that it cycles the forms of energy. The failure of the terms 'forward' and 'reverse' doesn't mean an end to a time direction, because even cyclical movement has a direction—the cycles can move either clockwise or counterclockwise, but we only observe the cycle moving in one such direction, not both.

Thus, time has a direction, but that direction is not an arrow! The direction is the clockwise or counterclockwise motion of a cycle, which is not reversible in the sense that it only moves in one direction. And yet, the irreversibility doesn't lead to a continuous increase in disorder (as in thermodynamics) because the old state is regenerated through the natural cycle.

Figure-33 Time's Arrow is a Cycle Direction

The directedness of cycles can be easily illustrated through examples. For example, a counterclockwise direction in a cycle would mean that nature prepares pizzas, cakes, and cookies, but we consume it as flour, milk, and sugar, and then we recycle the flour, milk, and sugar back to pizzas, cakes, and cookies. A clockwise direction in a cycle, on the other hand, would mean, that nature prepares flour, milk, and sugar, and we consume it as pizzas, cakes, and cookies. In principle, either of these alternatives could be true, but only one of them is. Therefore, after we reject the idea of time direction based on thermodynamics, we can accept another type of time direction that involves the recirculation of energy within an ecosystem.

The new problem is that the notion of cyclic time rests upon the existence of closed systems—i.e. ecosystems. For energy to circulate in cycles, there must be a boundary that demarcates one system from another. There are no theories of such boundaries in modern physics, although atomic theory is closer to accepting the reality of such boundaries than classical physics, because the states in a quantum system are mutually entangled, and changes to one state generally mean changes to all states in a system. Therefore, all principles of thermodynamics must be interpreted in quantum mechanical terms. First, as we have noted above, classical statistics pertains to the possibility of change in the quantum basis. Second, the arrow of time must be attributed to the circulation of energy across quantum systems (i.e. ensembles) aggregated into larger systems (i.e. ensemble ecosystems).

Hierarchical Space and Time

The problem of the direction of time leads us to the idea that nature is organized in closed systems, which cycle energy. We are intimately familiar with such systems in everyday life—economics, ecology, and even biology studies such cycles. In fact, there are cycles within cycles; for example, classical economics is conducted within an ecological cycle, and biological cycles exist within ecological cycles. Due to the embedding of closed systems within closed systems, we can envision a hierarchy of organization in which smaller boundaries are contained inside larger boundaries. This idea of organization is commonly applied to space in the everyday world; for example, the Earth is divided into countries, which are divided into states, which are divided into cities, which are divided into localities, and eventually into houses. Energy circulates in each of the regions and creates smaller ecosystems—e.g. a city recycles its waste (if it survives for a longer time). The smaller ecosystems then cycle energy with other similar sized systems, as part of a larger ecosystem. A smaller system also interacts with a larger system *through* its membership to another larger system. For example, while traveling, we invoke our citizenship to some country, residence in some state and city, etc. Even statistical mechanics is based on the idea of closed systems (called 'ensembles') as is the theory of atomic particles which are 'entangled' with other particles contained within an ensemble boundary.

The real question is: Should we attribute the structures of organization to *matter* or to *space*? If we say that matter is the cause of organized systems, while the underlying structure of space is flat and open, then we would have to explain how one possible distribution of matter is selected from infinite such distributions, and that, as we have discussed, is impossible. We earlier said that this problem can be resolved if causality is attributed to time, which will then select from a collection of preexisting possibilities. But since order emerges out of *selection*, therefore, order is neither created nor destroyed. When it seems created, then it has been selected. And when it seems destroyed, then it has been rejected, and another type of order which is relatively less ordered, has been selected. Thus, the problem of how order is created and destroyed in nature becomes completely irrelevant if we can describe nature as a space of preexisting possibilities to select from.

Thus, by careful consideration of numerous problems, we arrive at the conclusion that ecosystems are not *created* but *selected*. And what we call 'space' is not the world of observations but the world of possibilities. This space must have an innate *structure*—i.e. hierarchical organization that embeds smaller systems inside larger systems. If the world weren't organized in this way, and causality is attributed to matter, then there is no explanation of how order emerges from disorder. Thus, the idea of 'matter' is not separable from the idea of 'space'—if space is the domain of possibilities. This conclusion is consistent with atomic theory because entanglement of atomic particles can only be understood if the existence of a whole ensemble is recognized aside from the existence of the parts.

We have already dwelled on the causality involved in such closed systems: they compete to absorb smaller and larger systems, thus changing the notion of 'whole' and 'part'. Thus, sometimes, order seems to be created and sometimes it seems to be destroyed. But because the possibilities are eternal, therefore, the redrawing of system boundaries does not truly create or destroy order. They only select or access some preexisting order or reject such an objective reality.

Space-Time Interdependence

We now face an inevitable conclusion: time must be defined circularly as the appearance and disappearance of ecosystems, and this circular model of change itself depends on the ability to describe whole ecosystems. If ecosystems could not be described, then, time would not move circularly. And if time wasn't moving circularly, then the space of possibilities would have to be infinite, which would contradict the intuition that space is closed.

In short, the open and flat notions of time and space mutually depend on each other and mutually affirm each other. The open idea of time says that the past is never repeated, and therefore, the world doesn't exist as a possibility (possibilities are eternal and hence repeatable). The world is reality, which is constantly created from the present, as the past is destroyed. The idea that time is infinite now leads to the notion that space is also infinite, and in classical mechanics both were represented as flat, linear, and infinite containers. When the container contains reality, instead of possibility, then every change involves a destruction of the past, and causality in matter doesn't have a direction. Without a direction, the likelihood of the past repeating itself again and again is zero (even the possibility that the past would repeat itself even once is negligibly small). Thus, linear, and flat space and time are mutually affirming and mutually interdependent. The interdependence means that it is impossible to separate these two notions.

Similarly, the cyclical and closed notions of space and time mutually depend on each other and affirm each other. Since they depend on each other, therefore, they are the 'causes' of each other. And since they are affirmed by the other, therefore, they are the 'effects' of each other.

Ultimately, by 'inseparability' we mean that two things are mutually the causes and the effects of each other. They would be separable if they were only the causes or the effects. Mutual interdependence and affirmation of the ideas of space and time make them logically inseparable. This logical inseparability can also be understood as time and space not existing *without* each other. In short, there cannot be a static world of objects in which time doesn't exist and changes don't occur. And there cannot be a notion of time in which something passes but nothing changes as the evidence of that passing. In effect, the primary meaning of inseparability is that there cannot be a world existing in space without time, or a time that exists without a world.

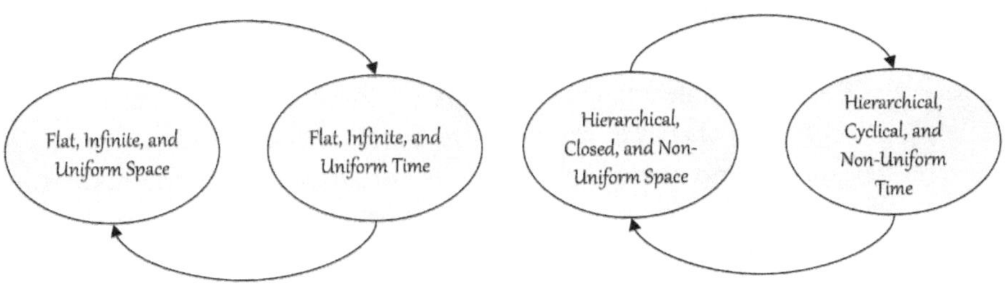

Figure-34 Logical Inseparability of Space and Time

If we change our ideas about space, then we must change our notions of time as well. If space is hierarchical, closed, and non-uniform, then time must have the same properties. Thus, the discussion of time becomes incomplete without discussing the nature of space, and vice versa. But this interdependence between space and time also helps us understand how this mutual causality and interrelation can be used for describing the emergence of the universe in which space expands time, and time expands space.

Takeaways from Irreversibility

The question of whether time is reversible, and the answer to this question, concludes this chapter. We can summarize these ideas as follows:

- The second law of thermodynamics talks about an arrow of time in which the universe goes from order to disorder. This arrow breaks the shortest-path principles, and the most likely state in the universe is where there is greater randomness than there is structure.

- James Maxwell conceived a 'demon' who could create order from disorder. His demonizing of order itself tells us what he was thinking, namely, that even if such a thing were possible, it would be evil. Since such things are indeed possible for conscious persons, their existence in a physical universe becomes the evil in the world.

- But we note how cellular molecular mechanisms themselves constitutes such 'demons' because they open the doors to ingest food and excrete waste and close the doors on toxins and attackers. But such molecular mechanisms could not exist outside of an ecosystem of cells in which 'waste' becomes 'food', and vice versa.

- Thus, after discussing why time is not an *arrow*, we also discuss why time has a *direction*. This direction is the cyclical circulation of matter in ecosystems because the ecosystem evolves only in one direction. Thus, the direction of time turns out to be the circulation of matter in cycles, rather than a linearly progressing 'arrow' of time.

- We then discussed the interrelation between space time concepts; namely, that flat, uniform, and continuous ideas of space and time mutually reinforce each other, while hierarchical, non-uniform, and discontinuous ideas of space and time reinforce each other. This mutual reinforcement leads us to a connection between space and time, namely, that they are mutually the causes and effects of each other. Effectively, time cannot exist without space, and space cannot exist without time. Hence there cannot be a static non-evolving universe, or an eternity of time-lessness without a real world.

- When time is defined due to a cyclical process, then the classical notions of time—e.g. as conceived as a straight line in Newton's physics and adopted in all subsequent theories of nature—disappear. The linear process is not the default process; the default process is a cyclical process. Escape from this cyclical process—if it ever occurs—is the linear process. This goes to the root of philosophical attitudes about the world where the observer is stuck in cycles, and an escape from these cycles can be considered progress.

8

Is the Universe Eternal or Cyclical?

Somewhere there lives a nameless world beyond the farthest star
whose sight escapes the searching gaze of watchers from afar.

The distant orb's inherent life emerged throughout the ages,
evolving into strange, exotic beings in languid stages.
—William Tatum

The Problem of Inertial Mass

Since the time of Newton's mechanics, the cosmic structure is supposed to be determined by a single property called 'mass', which exerts a gravitational force. While subsequent developments in physics revealed the presence of electromagnetic 'charges', due to the equal number of positively and negatively charged particles, which cancel each other's effects in any sizable system, electromagnetism has never had an important role in cosmological theories. Likewise, the strong and weak nuclear forces operate at such short distances that they only contribute to the stability or instability of the atomic nuclei and are not expected to have any influence beyond the nucleus. Thus, the forces responsible for the atomic structure of the atom—i.e. electromagnetism, strong, and weak forces—have no role in deciding the cosmic structure. They are, however, expected to play an important role in the early stages of the universe, when energy is highly condensed, because then, the subatomic particle interactions become important. Therefore, the focus of combining the effects of gravity with the other three quantum forces has been limited to studying such condensed matter-energy structures, colorfully called 'blackholes', 'white dwarfs', 'supernovae', etc.

Our modern understanding of gravity comes from Einstein's general theory of relativity, which is based on the classical physical equivalence between *inertial* and *gravitational* masses but extended using the constant speed of light. The inertial mass originated in Newton's laws of motion because accelerating observers experience a 'backpressure', which is not experienced by observers moving at constant speeds. In Newton's physics, this backpressure was attributed to a 'resistance' offered by the 'ether' in which the observers move. Galileo and Newton assumed that the amount of resistance offered by the 'ether' was proportional to the quantity of 'matter', and thus, the backpressure came to be identified as *inertial mass*. The gravitational force then overcomes this resistance and to compute the force exerted by gravity we needed another mass, which was called the *gravitational mass*. Since the gravitational force overcomes the inertia, and causes masses to accelerate, therefore, the inertial and gravitational masses were considered equivalent, and this subsequently came to be known as the *equivalence principle*.

The basis of the inertial mass—i.e. that the 'ether' exerts a backpressure—however, came under question after the advent of special relativity because the constant speed of light regardless of an observer's motion, seemed to imply that there wasn't an underlying medium through which light was moving. At least for the purposes of light, the medium did not seem to exist. Special relativity, however, was silent on the question of whether this medium existed for mass. Einstein's general theory of relativity makes claims that the ether doesn't exist even for mass. But if there is no underlying ether, then, how do we explain the origin of inertia? The general theory of relativity is silent on this question, and the reason is that the backpressure is felt by an accelerating observer, but this experience isn't seen by other observers, who only see the accelerating observer, rather than an observer who is accelerating *and* feeling a backpressure. By the principle of confining ourselves to 3rd person observations, the perceived backpressure is not relevant to scientific theories, because it only exists in the 1st person perspective. Just as we disregard the sensual and mental effects of physical properties, in the same way, inertia now becomes a private experience.

When these ideas are combined with the constant speed of light, the implication is that gravity doesn't act across space *instantaneously*. Rather, the gravitational pull travels at the speed of light, and therefore, the speed of light is the template for all forces. Thus, if a mass was moving at constant speed, the gravitational force will not take longer or shorter to affect that mass; just as light takes the same time to reach a stationary or moving object, similarly, gravitational pull will also take the same time. Thus, the general theory of relativity dispenses with the 'ether' completely on the premise that this principle can be applied both to gravity and electromagnetism. The indeterminism in classical physics, however, also extends into relativity[1].

The problem of inertial mass could be solved if we said that the observer is not *pulled* by the gravitational force exerted by another object. Rather, the observer accelerates toward the object of his own accord, and during this acceleration he *loses* rather than *gains* energy[2]. The loss of energy is experienced by the observer as inertia, while the acceleration is interpreted by the 3rd parties as gravitation. But that conclusion would entail that the law of gravitation is hubris, and no such law truly exists. Rather, it is possible to feel weightless even on earth, and without any special equipment.

The law of gravitation implies that the earth keeps us bound and prevents us from leaving the earth. The inversion of that idea could be that we bind ourselves to earth, but when we try to leave the earth, then our body resists it, and that resistance is experienced as the backward pressure. In short, what we call gravity of earth, is not a property of the earth, but a property of our body, and the resistance to leave the earth is not due to the earth applying some force on our body, but because the body itself applies a force against itself. If, for example, we were in a body that wasn't bound to the earth (due to itself), then it will not experience inertia, and we would not have to exert ourselves while trying to leave the earth's atmosphere.

This kind of approach will solve the problem of inertial mass, without sacrificing the equivalence principle, because the inertia of the body, and the exertion by the body, are both properties of the body, not of the earth. The exertion needed to leave the earth is equal to the binding to the earth. If the latter becomes zero, then the former also becomes zero, and then we experience weightlessness on earth, because the earth will not exert a force on the body; the body would be preventing itself from leaving the earth.

This scenario can be compared to a boatman rowing a boat tied to the shore. The boatman could think that the shore is pulling him backward, and he would be successful in leaving the shore if the forward force is sufficient to break the rope. But what if there were no rope binding the boat to the shore? Then, there would be no trouble in leaving the shore. Therefore, the problem of earth's gravity may not be attributed as a force between *two* masses; it can be attributed to a *third* entity—e.g. the rope—holding them together[3]. This type of causation would need to explain the third entity.

Inertia as Object Entanglement

To explicate the nature of this third entity, let us recall the discussion about semantics we had during the discussion of special relativity. I distinguished semantic information from physical information by calling them *hierarchical* and *linear,* respectively. A physically complex sentence is that which has many non-repeating letters; such a sentence cannot be 'compressed'. A semantically complex sentence also has a non-repeating letter sequence, but it also has an innate structure. We generally call this structure *grammar*. It includes a subject and an object connected by a verb; the subjects and objects are noun phrases, which in turn comprise other nouns, verbs, adjectives, conjunctions, etc. If we view a sentence as a sequence of meaningless physical entities, then physical and semantic complexities look identical. We see the meaning if we treat the words in a sentence as nouns that refer to things and their properties, and verbs that refer to actions.

A sentence, in fact, comprises three kinds of meanings. First, each word represents a concept. Second, these concepts describe the nature of things in the real world, and they are thus *references* to reality. Third, these concepts are organized in a grammatical structure that is expected to mimic the structure of reality. For example, when you say that "the sky is blue", the word 'sky' is a concept, which refers to the real sky, and it describes the relation between the sky and the property of the color being blue. The reality of the sentence is different from the reality of the sky, and either of them could exist without the other. The physical conception of reality disregards the existence of sentence-like objects. All objects therefore are like the sky, but they cannot be like the sentences about the sky. As a result, each of these three meanings are rejected. First, the objects are not the symbols of ideas. Second, no object refers to another object; each object is a thing-in-itself. Third, the relation between the things in the world is not reflected in the grammatical structure; indeed, there are only objects, and structure is an epiphenomena of their locations; there is no such thing as a grammatical structure in the real world which stands apart from the objects and their properties[4]. A grammar structure is distinct from the words, and the words are distinct from the things that they refer to. In ordinary language, we can distinguish between things, words (concepts), and structures. But the physical world just assumes that there are things, but no words or structures.

The problems of modern physics arise due to thinking of reality devoid of words and grammar. The problems of quantum mechanics, as we have seen, can be resolved if we treat quantum particles as *words* rather than *things*. And the conceptual issues of special relativity, as we have discussed, can be demystified if we think of measurements as exchanging messages, rather than particles. The natural extension of this thinking is that 'gravity' is not a force; it is a structure that binds objects, like grammar binds words. The atoms described by quantum theory are then words comprised of subatomic particles which are like the alphabets making up the words, and the gravitational force is like the grammatical structure that holds words together[5]. The grammar combines words by assigning each word a figure of speech; however, if we look at the words physically, we cannot see their conceptual reference, and we certainly cannot see the figures of speech. To understand the latter, we must view the objects themselves as symbols. And in that sense, the understanding of quantum reality is essential to solving the problem of gravity[6]. The additional insight from adopting a semantic view is that the grammar is not caused by the words, although the words are connected by grammar. In fact, the grammar is a separate tier of reality, which can never be perceived by the senses, although it can be grasped by the mind. Similarly, the objects exist in a structure, but the structure is not caused by objects. That structure is a separate type of reality distinct from the objects.

Thus, the 'force' of gravity can be described as a structural relation between the earth and our body. The earth doesn't bind our body; rather our body binds to a structure, which also binds to the earth. This is just like the shore that doesn't restrict the boat, although the boat ties to a rope, which is also tied to the shore. The structure also constitutes a 'space', but it is not the classical physical space, because 'positions' in this 'space' are like figures of speech—nouns, verbs, adjectives, conjunctions—in a sentence. To say that my body is bound to the Earth is simply to say that my body is like a word, that has accepted an additional state—a figure of speech, or a functional relation to the Earth, and my body is not the cause of that relationship. This leads to the conclusion that just as word can exist outside a sentence, similarly, my body can exist outside the relationship to the Earth. In that case, my body would not be bound to the Earth's gravity, and it could move freely, experiencing weightlessness. That idea contradicts gravity, and it depends upon saying that the effect of gravity is due to the structure.

In ordinary languages, we know that the same words can be used as a noun, verb, adjective, etc. We know that the same person can work in different roles in a society. And we know that chairs can sometimes be used as tables, and tables sometimes as beds. The physical composition of things does not change as we structure them through different relations. However, we assign additional attributes—a figure of speech to words in a sentence, a role to a person in a society, and a function to an object in our home. In a context, the figure of speech cannot be separated from the word. But in a generic sense, the same figure of speech can be attached to other words. Therefore, each word binds itself to a figure of speech, and hence to other words that are also binding themselves to the same grammar structure. However, since the grammar is independent of the words, therefore, the words are not bound *by* other words, but they are bound *to* other words. This is an important distinction because gravity rests on the premise that objects are bound by other objects—i.e. due to the masses in them.

The Rejection of Motion

The reason this problem seems so hard is because we started by assuming *motion*. We assumed that the distance between objects increases and decreases due to the motion of these objects. In short, the space that holds us is a physical entity, and the objects contained in this space are physical. This physical conception of space initially leads to the problems of length contraction and time dilation, as we have seen previously.

Subsequently, it leads to the total rejection of the experience of inertial backward pressure. These rejections are now stilted upon the idea that there is no absolute space, which begs the question: If everything exists in a reference frame attached to an object, are we not treating the object in two distinct ways—(1) as a material entity, and (2) as a space-time—which in effect creates the conceptual problem that the same object exists in many space-times, pertaining to each observer, so we must be breaking objectivity in some way. We could for example say that there is an objective reality in some objective space-time, but relativity cannot describe this objective reality, and that would be the correct sense in which the theory is incomplete. However, relativity takes the opposite position to says that a copy of each object exists in each reference frame tied to each observer, therefore, everyone has a private clone of the world, and yet, we are some how able to *communicate* with each other, which involves an information transfer *between* such space-times.

The paradox of relativity is simply this: If the relative space-times are indeed the only reality, then how does information go from one space-time (object or observer) to another space-time (object or observer), when information must be *within* space-time and not cross over to another space-time? If indeed information crosses over from one space-time to another, shouldn't there be another space-time through which it passes? If not, then what is the mechanism by which information goes from object to object? Unless we solve this problem, relativity is conceptually problematic.

These conceptual problems do not arise if we think of the world as possibilities, because then there is no motion. As we saw earlier, proximity is simply a stronger interaction between possibilities. If the strength of interaction increases linearly, then we can call that constant speed motion. If the strength of interaction increases non-linearly, then we can call that acceleration. All these interactions occur in an absolute space, so nothing is going 'outside' space, or 'in between' spaces. Our basic intuition about space is that things happen within space, and nothing goes outside space. Relativity breaks those intuitions to explain the observers with the postulate of a physical world, a physical space, and a physical change—i.e. motion. Therefore, we must reject the ideas of physical world, space, and change. The space of possibilities is a solution that explains all the observations, without breaking the basic conceptual intuition of the world being *in space*.

However, science doesn't work merely by conceptual intuitions. We also must provide empirical evidence for preferring one viewpoint over another. Fortunately, the alternative view of an absolute space time provides exceptions to relativistic motion. As we noted above, an object can seem to accelerate toward another if the strength of interaction increases non-linearly. This is an exception to relativity because in this case, despite acceleration, no backpressure would be felt. We can call it *inertial* acceleration. Similarly, another exception to relativity is the existence of *non-inertial motion*—in which changes to distance represent the changes to structure. This type of change can also create the appearances of constant speed or acceleration (although there is no physical motion). Therefore, the exception to relativity in this case would be the presence of a *non-inertial constant speed* motion.

Classical physics and gravitational theory equate inertial motion with constant speeds, and non-inertial motion with acceleration. However, if we solve the conceptual problems of relativity, then we arrive at two kinds of differences with relativity—some constant speed motion can be non-inertial, and some acceleration can be inertial. In the first case, even though the distance is decreasing linearly, a backpressure would be experienced. And in the second case, even though the distance is decreasing non-linearly, no backpressure would be experienced. The postulate of an absolute space is therefore not merely conceptually convenient, but it also makes additional predictions, which are contradictory to those of gravitational theory.

Finally, the possibility absolute space also tells us that the amount of backpressure felt during a structural change must be related to the size of the structure. The larger the structure, the more resistance it will offer to changes, and less connected structures will exhibit lower inertia. The result would be that different levels of inertia would be experienced in different cases—depending on the extent of connectedness of the structures. This will lead to the dismal conclusion that gravity is not uniform across the universe. If gravity is not uniform, then it cannot be a law of nature. But the failure of the gravitational law can be explained using structural ideas.

Thus, the structural view of mass presents not merely a conceptual convenience, but empirically confirmable effects: (1) all accelerations are not non-inertial, (2) all constant speed motions are not inertial, and (3) gravitational effect is not uniform. Obviously, if there were no counter-indications to gravity, then these conflicts with gravity would immediately disprove the structural idea of gravity. But the fact is that there exist counterindications, which are called 'dark energy' and 'dark matter'. It is estimated that only 4% of the universe obeys gravity; the remaining 96% of the universe contradicts gravity. Does that make the structural view true? Not necessarily. By realizing that gravity fails in 96% of the cases, we merely prove that gravity is not a universal law; that it seems to work in some cases, but it cannot be true. Then, to prove that the structural alternative is true, we also need to demonstrate that the observations are consistent with its other two predictions—i.e., (1) inertial accelerations, (2) non-inertial constant speed motions.

Dark Matter and Dark Energy

Let's therefore turn to the experiments that have led to the hypothesis of dark matter and dark energy. The first key observation is that the rotational speeds of the distant parts of galaxies do not conform to gravity (the gravitational theory only applies to the innermost regions of a galaxy). According to gravity's predictions, the farther regions of galaxies must rotate much slower than the inner regions, because the gravitational pull on the outer regions decreases, and with the reduced inward pull to these outer regions of the galaxy, the outer regions must also rotate slowly. A simple example of this principle is the Solar system in which the other planets such as Jupiter and Saturn rotate much slower than the inner planets such as Mercury and Venus. A galaxy is also like the Solar system in the sense that the *visible* mass of the galaxy is concentrated in the center, and if we take this visible mass to be all the mass in the galaxy, then the galactic rotations must also be similar to the rotations seen in the Solar system[7]. The observations, however, reveal that once we go beyond the innermost regions of the galaxy, instead of slowing down, the outer regions of galaxies rotate at a constant speed. This is an example of *inertial acceleration*—the rotation is acceleration, but it doesn't follow the principle that the farther regions of the galaxy must rotate slower, and therefore, it cannot be termed a non-inertial acceleration. The constant speed of galactic rotations violates the predictions of gravity, and the violation is that in some cases, accelerations seem to be inertial.

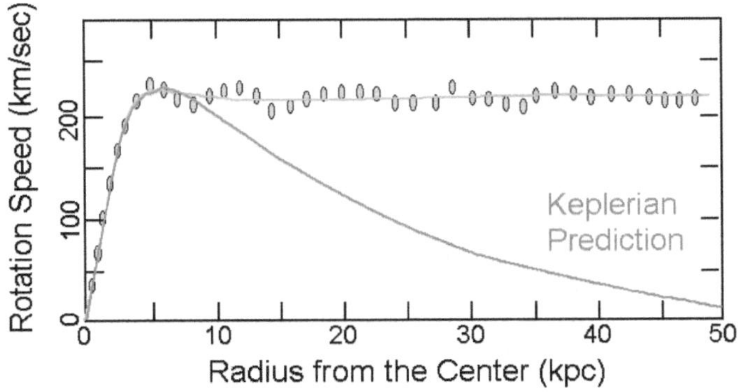

Figure-35 Anomalous Rotation Speeds of Galaxies

The second violation of the gravitation theory is the so-called Cosmic Microwave Background or CMB. Since the universe is uniform in matter according to relativity, all regions of the universe must essentially produce the same light spectrum—i.e. the brightness of the spectrum can decrease with increasing distance to another galaxy, but the frequencies of the light received from other galaxies must be the same. Observations, however, show that the light from the other galaxies is redshifted—i.e. the intensities of light commensurate with the distance are seen on lower frequencies. This fact cannot be explained by saying that the universe is expanding at a constant speed, because the speed of light is constant across such moving reference frames. It can only be explained if we say that the universal expansion is *accelerating*, because in an accelerated expansion universe, time will indeed take longer to travel, and that longer time translates into a lower frequency, which would then be perceived as the redshift of the light spectrum.

When Einstein formulated general relativity, he did not know if the universe was contracting, expanding, or static. All the observations until that time seemed to suggest that the universe was static. Since the Big Bang causes the universe to rapidly expand, to counterbalance this expansion (and produce a static universe) there must be a force that controls the expansion. Of course, gravity could be that force since it is attractive. However, it wasn't obvious if there was enough matter in the universe to control the expansion. This 'just enough' mass is called 'critical density' or $\Omega 0$. If the total mass in the universe is $\Omega 0$ then the universe will stop expanding after some time, and its size at that time would remain fixed forever. If the actual mass is larger than $\Omega 0$ then the mass will halt the expansion and the universe will go back to Big Crunch. If the actual mass is lesser than $\Omega 0$ then the mass will not halt the expansion and the universe will expand forever, although the expansion will get slower over time. At that time, however, nobody had anticipated that the expansion of the universe could *accelerate* over time.

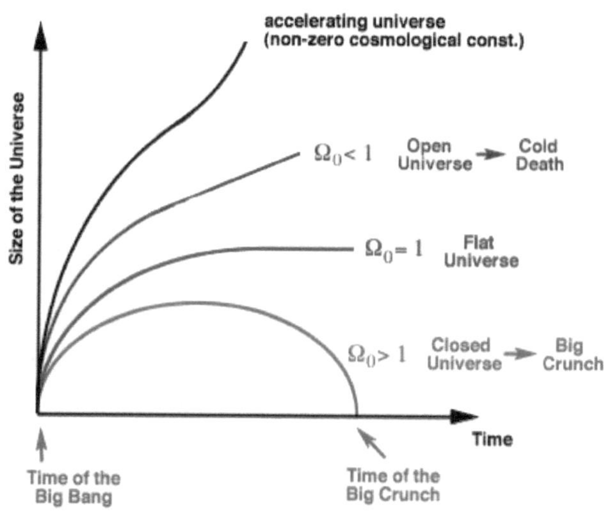

Figure-36 Anomalous Accelerating Expansion of the Universe

In 1998, redshift measurements on some galaxies revealed that rather than slowing down, the galaxies were accelerating, implying that $\Omega 0$ is negative. To explain the acceleration, physicists postulate the existence of "dark energy" which works like anti-gravity: i.e. it pushes matter apart, and with increasing distance, the anti-gravitational force increases, thus accelerating the expansion. Since anti-gravity is required to explain the observations, therefore, the observations contradict the universality of gravitation. This is now called the Big Rip, which will tear the universe apart. The redshift values indicate that "dark energy" must about 68.3% of the universe.

Since light moves at a constant speed, but light is also redshifted, therefore, this is an example of *non-inertial constant speed* motion. The constant speed pertains to the light's motion, and non-inertia pertains to the accelerated expansion of space. A constant rate of expansion would go undetected because light would behave as if this was another inertial motion, and a constant rate of universal expansion would go undetected. There are other ways to solve this problem, which don't involve an expansion. For example, we can say that the other regions of the universe are not uniform in the properties of quantum particles, and so they emit a different spectrum. Likewise, we can say that the remote galaxies don't interact with the present galaxy in the same way as the regions within this galaxy do. All such solutions, however, would violate the truth of quantum theory, instead of relativity. To an extent, it becomes a choice for scientific theory formation about which theory we want to change. But to the extent that we assume that the quantum particles are the same everywhere, and they always form the same type of Hydrogen and Helium atoms, it is a more compelling argument to say that the universe seems to be expanding[8]. That expansion, however, needs the presence of anti-gravity, and contradicts gravity.

Thus, we not only see the breakdown of gravity, but also see that this breakdown occurs in two specific ways— (1) inertial acceleration, and (2) non-inertial constant speeds. According to gravitational theory, all acceleration must be non-inertial, and all constant speeds must be inertial. The equivalence of inertial and gravitational masses rests upon this principle. But if the principle is violated, then gravity itself is not a universal theory.

Three Kinds of Space and Time

On that basis we can say that the basic idea underlying general relativity, namely, that there is no underlying space, is false. Moreover, to reconcile quantum theory with relativity, we must distinguish between the space of fundamental particles (they must be understood as fundamental symbols of meaning, with 'distance' between them described as the difference in meaning) from the space in which these particles are organized into functional structures. Once we recognize the existence of meanings, then we must also distinguish between the *meaning* and its *instance*—i.e. a symbol. What is an instance? Recall that we describe matter too as a type of consciousness, although different from our consciousness. We also said that matter and observer exert a mutual responsibility on each other. To exert this responsibility, each of them must be treated as *persons*. The instance of an idea is a person with intentions, with abilities—that we can call their 'body' that encodes meaning. The combination of meaning and instance is therefore the combination between meaning and intention. The meaning represents properties such as color, size, form, etc. and the intention is the purpose for which these properties are to be used. In modern science, we are used to physical instances; for example, a particle is the instance of the idea of a particle (i.e. something that moves at constant speed unless disturbed by a force). Thus, both the instance and the properties are physical. In the semantic view that I describe here, we must treat both property and instance semantically—the property is meanings, and the instance is a purpose.

These distinctions between meaning, intent, and structure, lead us to three kinds of spaces. First, there is a space of semantic possibilities in which the distance between locations represents conceptual difference. Second, there is a space of intentions that instantiate these concepts into symbols. Third, there is a structural space that combines the *instances* of these concepts—i.e. symbols—into functional structures. These three kinds of spaces are simplistically illustrated in Figure-37. The simplification is that each of these spaces must be treated hierarchically, because otherwise we will have no way of counting or numbering the locations and entities in these spaces.

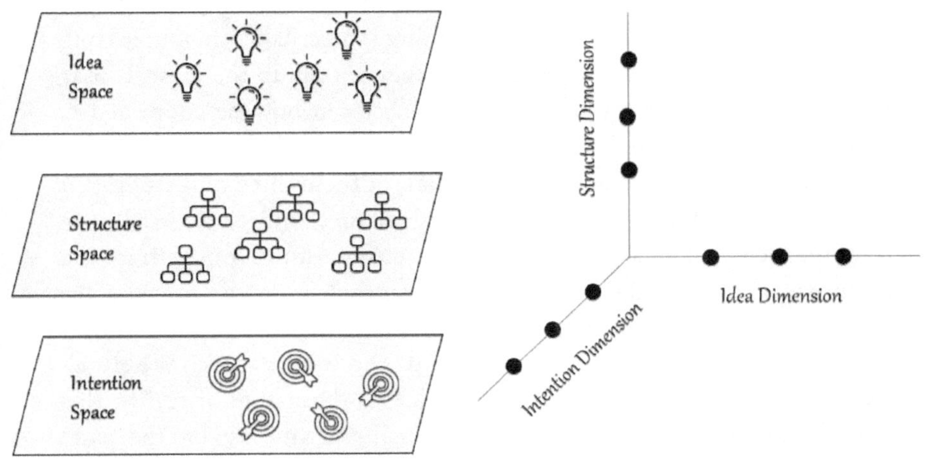

Figure-37 Three Kinds of Spaces and Dimensions

Since the entities in each space can be numbered or counted, there-fore, each such space can be described as a *dimension*, and these three kinds of dimensions would now constitute a three-dimensional space, although the dimensions will no longer be physical; they will be semantic dimensions. Each of these dimensions also constitutes a different kind of possibility, such that an observable in this three-dimensional space must combine the three possibilities. Thus, we can say that each observation has an idea, individual, and structural dimension, and the possibilities in each dimension are known only when the dimensions are combined. In short, the dimensions are the possibilities, the origin of the space is the point from which the universe of possibilities springs, and the obser-vations are the locations in this space (although they are not locations along the possibility dimensions). The observable world thus excludes the dimensions which must be considered the objective reality prior to observation. These possibilities are not experienced unless the possibili-ties along the three dimensions are combined.

The structure that combines the instances of meanings is the result of effects and consequences produced from previous choices. So, even though the effects and consequences are possibilities at the present, they are in some sense, the products of the past. The intentions that individ-uate meaning similarly exist in the present, but they refer to the future.

Finally, the meaning exists at the present, and it constitutes the current state of the system. Thus, the state of reality is described using three components—that exist in the present, but they pertain to the present, past, and the future. The evolution of this reality can now be connected to the nature of time, and corresponding to the three dimensions in space, we can describe time too as the evolution of these three dimensions. As choice, time selects some individuals, structures, and concepts. But in another sense, time activates some goals in the future, some effects and consequences from the past, and some capabilities in the present. The *causality* in time can therefore be described in terms of these three dimensions, which can be called 'past', 'present' and 'future' which are simultaneously existing in the present, but due to the nature of reality, that combines these three dimensions, the causality in time also has the 'past', 'present' and 'future'—all existing in the present.

Time is not one-dimensional while space is three-dimensional. Rather, both space and time now become three-dimensional entities; the difference between them would simply be the difference between possibility and choice. In a sense, time 'evolves' the visible space by making some possibilities feasible at a given time, while other possibilities remain hidden[9]. Thus, we can make two kinds of novel statements about time. First, the past, present, and future—as possibilities—are eternal; these possibilities constitute space. Second, the past, present, and future, as the dimensions of time that activate consequences, abilities, and intentions, exist at the present. According to the first, the past, present, and future are *eternal,* and according to the second, the past, present, and future exist in the *present.* The eternity is about the possibilities of space, and the present is about the nature of time.

Time, however, doesn't combine these three spaces, because that combination would lead to the denial of free will, and the universe would become deterministic. The combination of the three dimensions, to produce a specific observation, can now be attributed to the observer. Each observer can combine some idea, with some structure, with some individual, and that combination represents the observer's *choice* and *experience.* Since choices entail responsibilities, and responsibilities entail that possibilities available to an observer can expand, contract, or change, therefore, each observer is restricted in terms of which of the possibilities along the three dimensions he can combine. Because of the effects and consequences of previous choices, an observer will only have access to a proper subset of the possibilities.

In simple everyday terms, we can illustrate this by an example. At a given time, certain models of vehicles will appear as ideas. At that time, there will be some individual entities that instantiate that model or design of the vehicle, and thus constitute a *symbol* of the idea. Such vehicles will exist in some market of producers and consumers of the vehicle. According to the effects and consequences of an observer's choices, some observer will get an opportunity to produce such vehicles for that market. Similarly, based on the effects and consequences of their choices, some observer will get access to this market and would be able to acquire a vehicle. The observers are, of course, not compelled to produce or consume such vehicles, because the combination of possibilities is a choice; therefore, the participation in the market of vehicles is a choice. However, choices are not unlimited; they are constrained by the effects and consequences of previous choices. Thus, some observers get access to the market, and choose to buy a vehicle. Other observers get access to the market but choose not to buy the vehicle. Some other observers do not get the access to the market of vehicles. And even other observers are compelled by circumstances to buy the vehicle. If they get the access to a market, and buy that vehicle, then they get a *result*. This result is the ownership of the chosen vehicle, and it is produced due to the combined effects of— (1) three kinds of spaces, (2) three kinds of times, and (3) three kinds of causalities—effects, consequences, and choices.

Since the possibilities are not created or destroyed, therefore, matter and space are eternal. Since time selects a subset of these possibilities, and the space of possibilities is finite (recall that choice entails responsibility only in a finite universe), therefore, time must cycle through these possibilities and this cyclical time is also eternal because there is no limit to how many times we can iterate over a set of possibilities. Finally, since the observer must reap the effects and consequences of their choices, therefore, even the observer must also be eternal. Thus, space, time, and observer are all eternal in this picture of reality. The temporariness of the world pertains to— (1) the choices, (2) the effects and consequences of choices, and (3) the selection of the possibilities from the set of all the possibilities due to time.

Since the choices, effects, consequences, and the results are temporary, therefore, despite the eternity of the observer, the observer's experience is temporary. Similarly, despite the eternity of both space and time, their combined effect is temporary. In one sense, the ingredients that produce the changing world are eternal, and yet, the observed world is temporary.

This temporariness is both objective and subjective because the world is objectively changing due to the combination of space and time, and subjectively evolving due to the choices, and their results, effects, and consequences. The eternity of space, time, and the observer entails that we can construct causal explanations based upon a conception of reality. And the changing nature of their combinations implies that this explanation can be used to do predictions. Finally, due to the laws of choice—which are mutually exerted by space, time, and the observer upon each other—because these persons (we identified them earlier as Nature, God, and the soul previously) this prediction can also be rational and empirical. Thus, we satisfy all the conditions for a scientific theory— (1) there is an objective conception of reality, (2) there are reliable predictions of observations, and (3) these predictions are explained using a rational theory of the reality.

The Process of Choice

When we start considering the combinations of ideas, structures, and individuals, then we run into a serious problem—these possibilities are not always *compatible*. As an example, some words or sentences cannot be meaningfully said in some situations. Sometimes, we don't want to say certain things in certain situations. And sometimes, the situation forces us to say something that we don't want to. As a result, the combination of these three possibilities produces many incompatibilities, and choice then resolves these incompatibilities, by making a *compromise*—all that is to our liking is not necessarily combined, and all that is dislikable must sometimes be combined. The essence of choice is therefore not merely a selection, but the combination also entails conflict resolution for it to remain meaningful. That meaning can be universal—i.e. somethings are always meaningful; it can be contextual—i.e. that which is only meaningful in a context; and it can be individual—i.e. that somethings are meaningful to us, but not others.

Pursuant to this model of choosing, we can refine our understanding regarding the process of possibility combining. We can discern the following stages through which a pure possibility becomes an experience:

- There is a realm of pure possibilities which exist eternally, but may not always be accessed all the time, or by everyone. We have called this the eternal domain of possibilities, which are conceptual.
- Time selects some of these possibilities and they become feasible at a certain time, at which point, they can also be perceived.
- The accessing of the temporal feasibility creates incompatibilities between the feasibility, and our abilities from the past, the goals in the future, and the exigent circumstances at the present. Choice is now required to resolve the incompatibility through a compromise.
- Once the incompatibility is resolved, the goals are modified, and a new trajectory to the goal must be determined to reach the goal. This trajectory can be called a 'plan' used to obtain the goal.
- The plan determined in the previous step must be executed, and that execution of the plan requires energy and effort from us.

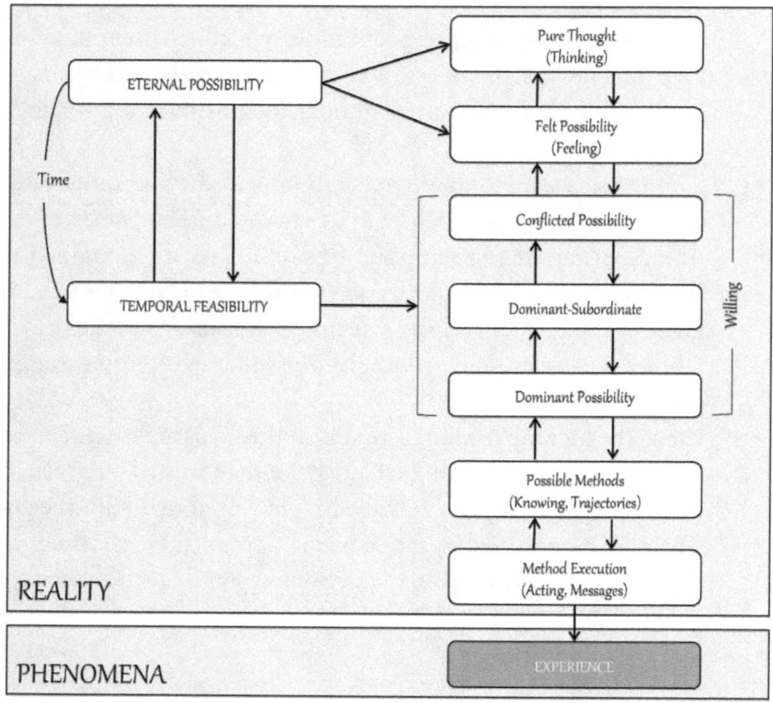

Figure-38 Conversion of Possibility to Reality

Space comprises the eternal possibilities so everything can be conceived and felt at any time. The eternal possibility sometimes enters our minds as unprovoked thoughts; these should not be thought of as our mind's creation, but an interaction between the mind and that possibility. Similarly, feelings automatically arise in us without an external interaction; these too are not our mind's creation, but an interaction between the mind and an eternal reality. Thus, the eternal possibility can be conceived and felt eternally—i.e. we can understand, and like or dislike, this eternal possibility. However, we cannot eternally perceive this possibility via our senses.

Time converts the eternal possibility into perceivability—i.e. the possibility becomes *feasibility*. Once we see that something is feasible, then we undergo a process of conflict resolution about whether something is available for us, whether it will be good for us, and whether it can be obtained rightfully. Due to this process of conflict resolution, which can be called 'willing', we arrive at a decision about what we want to do, or what we can call our *goal.* To reach this goal, we need a *trajectory*, which is the process or method by which the goal can be attained. Finally, we must walk this trajectory and act upon the process. The process of choice is therefore the process of *focusing* our attention from infinite theoretical possibilities (thinking), to those possibilities which we like (feeling), to those possibilities which we determine are best for us (willing), to the method of attainment of these goals (knowing), finally to the actions of interacting with the world through informational messages with the other observers or objects (acting).

The interaction now produces an experience that has been consciously understood, desired, judged to be suitable, planned, and executed. This type of experience constitutes what we call 'state preparation' and 'observation' in science. It is different from what might randomly enter our minds as ideas, or the feelings that might suddenly crop up, or things that we might see accidentally, or actions that we might perform inadvertently. If we understand the process of scientific experimentation, then we can also explain the other inadvertent processes through a similar method. For example, things that enter our minds as idle thoughts may be disliked immediately; in this case, the feeling terminates the thinking. Similarly, if some thoughts are liked, but are judged to be inappropriate, infeasible, or not good for us, then the judging terminates the feeling and thinking. Likewise, things that are thought, liked, and judged to be feasible, appropriate, and good for us, might turn out to be too difficult to attain; the thinking, feeling, and judgment are therefore terminated at the planning stage. Finally, we might produce a plan for attaining what we have thought, liked, and judged to be feasible, appropriate, and good, but we might not have the capacity or ability to execute; the previous stages of thinking, feeling, judging, and planning are therefore terminated at the execution stage. Ultimately, what we perceive by a conscious process is the outcome of satisfying these five conditions and can be called 'conscious activity'. There are other activities that might be done inadvertently, unconsciously, due to wrong judgments, or without an execution plan, and they are—to the extent that some of the steps are absent or weaker—less conscious, deliberate, or intended actions. If we understand the process by which deliberate, conscious, and intended actions occur, then we can remove or weaken these steps and understand the unconscious actions. For example, in an inanimate process, there is no thinking, feeling, judging, or planning. There are simply actions, which might follow the thinking, feeling, judging, and planning of a conscious observer. The operation of a machine, in the hands of an operator, for instance, follows this process. The operator does the thinking, feeling, judging, planning, and some execution—e.g. pushing buttons or pulling levers—and the machine does most of the execution. We cannot say that the actions of the machine are happening 'automatically', because there was a considerable amount of conscious processing that went inside the observer. Just because they are missing in the machine doesn't mean that they are missing in the world.

When modern science describes the world as a machine, it considers only the execution, and the planning, judging, feeling, and thinking processes are ignored. This execution seems like a physical process because the deeper levels of reality that lead to it are ignored. Now, even the machine is modeled as not executing a plan, but simply pushing and pulling. And with the successes of the machines that effectively push and pull, a false caricature of nature is produced by science—everything is this push and pull—and the processes of thinking, feeling, judging, and planning are supposed to be produced by such push and pull, although we cannot factually explain that reduction, because the underlying mechanisms are those of choice.

Five Kinds of Actions

In our previous discussion, we noted the 'actions' of transacting messages to create an observation, which is the step just prior to experience in the picture above. We also noted above that the interactions can produce the appearance of stationary position, constant speed motion, and acceleration. Therefore, two objects that seem to be moving or accelerating are not moving or accelerating; they are interacting, and the strength of interaction determines the proximity. If the strength of interaction grows linearly, then we can call it uniform motion. If the strength of interaction grows non-linearly, then we can call it acceleration. Thus, stationary states, constant rate motion, and acceleration are the effects of different types of interactions. More precisely, such differences must be attributed to the differences in the information exchanged between the observers and the objects.

Let us now turn to the understanding of the 'action' stage, based on the idea that it leads to experience. What is experience? It constitutes two things—knowledge and activity. By knowledge we grasp the nature of things, and through our activity we change the world. The process of knowledge, as we have discussed, requires the sender *cloning* its state, and the receiver *comparing* its state to the state encoded in the clone. Similarly, the process of activity requires the sender issuing a *command*, and the receiver *executing* this command. In each case, there is a source and a destination of a message, and the message either carries either knowledge or command. These messages also have the property of 'strength' and 'power' as we have discussed, so they are not merely physical particles, but symbols of information, and the strengths and powers create the contrast and brightness, which is then perceived as the object being near or far. By varying the 'strength' and 'power', the stationary, constant speed motion, and accelerated motions can be indicated in the messages exchanged between the observers and/or objects, which will then alter the causal effects. Now, stationary state, constant speed motion, or accelerated motions are properties of the information exchange, not the properties of factual motion.

In a simple sense, while performing an observation or action, the source of information must indicate their state (i.e. stationary state, constant speed motion, or accelerated motion), which is carried to the destination. The destination executes the requests and commands, after considering the state in the request or command, issues a result, which also describes its own state (i.e. stationary state, constant speed motion, or accelerated motion). Thus, through informational exchanges, the source and destinations of information know each other's states, and that information then produces the observed effects. For example, the observed effects of motion would be that temperatures will seem to rise, and masses will increase.

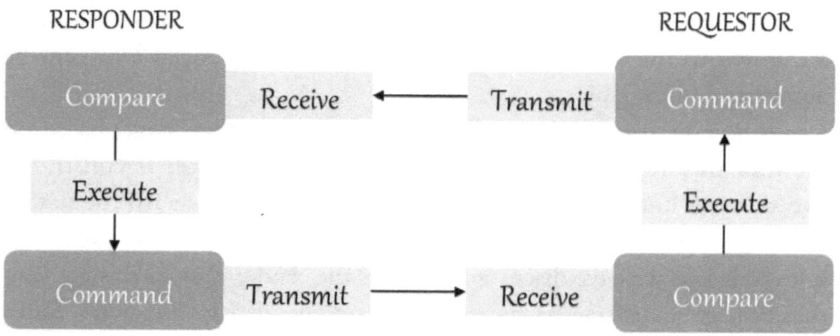

Figure-39 Five Action State Machine

To describe the informational method of producing observed changes, we can update the simple Request-Response mode of information exchange that we earlier depicted in Figure-15. This model must describe both knowledge and activity. The model of interaction relies purely on the types of messages, and the contents of these messages. Below is a summary[10]:

- Command: This is the action requested by the sender from the receiver. This command is modified by a moving or accelerating object, and therefore, it also encodes its current state information.
- Transmit: This is the action of sending a message to a receiver. It involves the choice of selecting to send to a certain receiver.
- Receive: This is the action of receiving a message from a sender. It involves the choice of selecting a sender to receive from.
- Compare: this represents the assessment of what the next state must be based on the received command. The next state can be the current state if the command simply requests information.
- Execute: this represents the change of state in case of an activity. It involves the transition from one state to another. Since all states are preexisting, the execution is the choice of the next state.

All commands comprise of three things— (1) the sender and receiver of a message, (2) the action requested by the sender from the receiver, and (3) the state of the sender requesting an action from the receiver. The action can be the request for the receiver's state, or an instruction to change its state. And the sender issuing a command must indicate its own state—i.e. if it is stationary, in a state of constant speed motion, or acceleration.

These states are believed to be unknowable if the 'ether' doesn't exist. In special relativity, it is impossible to know if an observer is stationary or in motion, because the motion is relative to other objects. Likewise, in general relativity, it is impossible to know if an observer is accelerating because the experienced force could be due to inertial or gravitational masses.

However, if all observers are always stationary, then the state of motion doesn't exist, but it is still possible to speak about 'stationary', 'constant speed' and 'acceleration' because these are the *chosen* states; we cannot not know what we are choosing, so an observer always knows the state, and that state can therefore be indicated in the messages that are sent or received. When this information is indicated in the requestor commands, then the receiver also learns about the requestor's states and it knows its own state. Therefore, during each message exchange, each side learns about the other object's states, and it knows its own state. Hence, the central premise of relativity, namely, that we don't know our state, and we cannot know the other observer's state can be removed via information. A theory of nature should not be built on supposed ignorance about the world. A theory should rather be built on the premise of perfect information, although if this information is false, wrong, or bad, then the choices are *irresponsible*, and they lawfully entail the effects and consequences of irresponsible choices.

The Origin of the Universe

During the discussion of relativity, and its causal explanation, we had noted that as contrast increases, the brightness declines, and vice versa. We also noted subsequently that there must be a smallest unit of information. This smallest unit of information entails that as an observer gets closer to an object, the contrast will continuously increase, but up to a point. Beyond that point the brightness becomes zero, and that zero brightness entails the absence of an informational transaction. Without a transaction nothing can be perceived. Therefore, two possibilities (in the possibility space) can be interacting with infinite strength, but they would be invisible, because despite the proximity, the brightness is zero. As a result, this state of the universe, when all the possibilities are interacting strongly would also be an invisible universe. Conversely, as the strength of an interaction declines, the brightness increases, along with the distance to the object. But the increase in distance is limited by the size of the universe—remember we are talking about a closed space with a boundary, not something infinitely extended (choice exists with a responsibility only within a finite universe). Therefore, as we move farther from material possibilities, until some point we can see them because their brightness is also increasing proportionately. But as we move even farther, we also see more and more abstract meanings, and the messages from the distance carry these meanings. If we reach the limits of abstraction, then there is nothing meaningful left to be said. Now, two objects are infinitely bright, but nothing would be visible because without meaning there is no information exchange. Thus, a universe, where all the possibilities are interacting weakly would also be an invisible universe.

It is worth remembering again that the world is not expanding or contracting, but by changes in the strengths and powers of interaction, they *seem* to be near and far, expanding or contracting. Upon this basis, we can describe the origin of the world in a new way: the world always exists as a primordial possibility, but nothing is visible if the possibility interaction strength is infinite, because despite zero distance, the brightness is also zero. The world becomes visible as the interaction weakens, the contrast weakens, and brightness increases, and the vision of many individual things is created. Conversely, as the strength of the interactions decreases, the brightness also must go up, and the world disappears from our vision, because despite infinite brightness, the contrast is zero. Now the question arises: Why would contrast and brightness change their strength and power? What causality can be the cause of the appearance and disappearance of the universe?

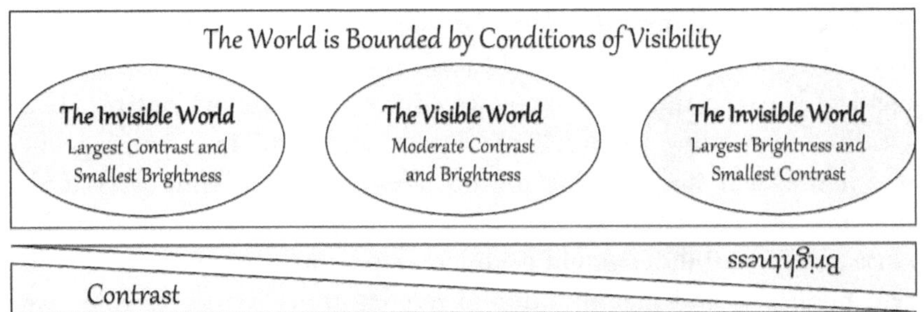

Figure-40 Perception Determines the Bounds of the World

We can recall that only one of these parameters—strength or power—need to be changed to change the other parameter. Therefore, the universe can appear or disappear if we simply change either the brightness or the contrast. Here, we can consider the appearance and disappearance of the universe because of the change in brightness; it is the energy that carries meaning. If all information transactions become high-energy, then the universe will seem to grow farther apart and appear to be 'expanding'—although there is no real expansion. Likewise, if all information transaction become low-energy, then the universe will seem to come closer together and appear to be 'contracting'—although there is no real contraction.

Thus, the universe can seem to contract and expand by changing the power of the interactions; in short, if the same meaning is transacted with decreasing energy, then the universe seems to collapse; and if the same meaning is transacted with increasing amounts of energy, then the universe seems to expand. In a linear universe, the universe could either begin with infinite brightness and zero contrast, and begin contracting, or it could begin with infinite contrast and zero brightness and begin expanding. A cyclical universe, on the other hand, would toggle between these extremes, and sometimes seem to expand from nothing (the supposed 'Big Bang') and sometimes seem to contract to nothing (the supposed 'Big Crunch').

Therefore, whether we talk about a linear or a cyclical universe, the mechanisms can be the same; the difference would simply be that the universe will have a finite (although very long) existence in a linear model, and an eternal existence in the cyclical model. A linear universe would either head toward a 'Heat Death' (due to infinite brightness), or a 'Cold Death' (due to zero brightness). A cyclical universe will instead oscillate between these two extremes—sometimes coming out of 'Heat Death' and heading toward 'Cold Death' and sometimes coming out of 'Cold Death' and heading toward 'Heat Death'. If time is cyclical, then the universe would go from invisible to visible to invisible, but the *alternating* invisibilities would be different: one invisibility will be due to zero brightness and the other due to zero contrast. With everything we have discussed thus far, the cyclical universe is more consistent, although we cannot determine its nature from within the universe, because time causes the universe's appearance and disappearance. From within the universe, both models seem equivalent, however, if we remember all the other cyclical processes, then the chances of the universe being cyclical are overwhelmingly higher than a linear one.

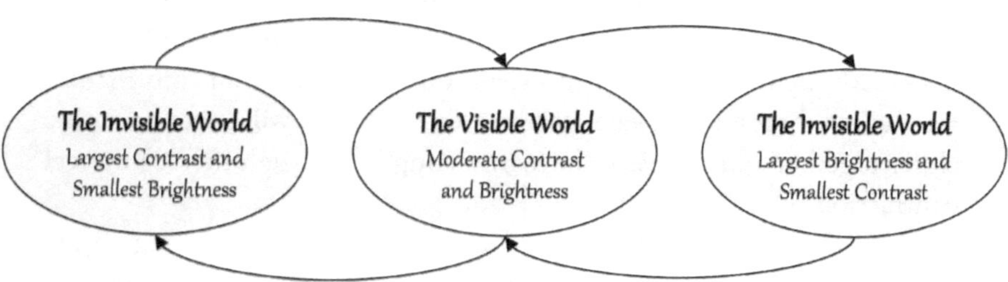

Figure-41 The Cycling of the Visible and Invisible Universe

However, these are only *phenomena*. When the world is visible, it has not 'appeared', and when it is invisible, it has not 'disappeared'. The "big bang" and "big crunch" are phenomena of apparent or phenomenal spacetime. During the big bang, the interaction between the possibilities begins to weaken, and the things *seem* to move apart, and the universe gets brighter although more obscure, creating the illusion of an expanding universe. During the big crunch, the interaction between the possibilities begins to strengthen, and things *seem* to come closer, and the universe seems to get clearer although dimmer, creating the illusion of a collapsing universe.

Using this idea, we can say that what modern science calls the red-shift of galaxies (attributed to the so-called expansion of the universe), is due to the weakening of the strength, which creates the appearance of expansion, without a real expansion. It is one of the two possible scenarios for the universe, and since they alternate, therefore, we are presently in the expansion scenario. In the next iteration of the universe, if we are still around (remember that the observer is eternal), we would be in the contracting scenario. There can be overwhelming theoretical evidence for a cyclical universe, but there can be no empirical evidence for it, because all the empirical data is wiped out when everything becomes invisible. Hence, we cannot infer from any observation about the cyclical nature of the universe, but if we form the theory of the universe based on the principles that we have discussed above, then we can conclude that the universe is also cyclical.

The Causality in Time

While discussing the Self-Moved Mover, I had described why it is important to postulate a self-caused change—(1) the alternatives of a time-particle moving on a time-axis necessitates an infinite cascade of times, and (2) if causality is imputed upon matter, then we require infinite amounts of energy at the beginning of the universe to create matter, and at every moment to compute the laws of motion. We subsequently discussed the problem of relativistic time, and its solution, namely, that perceived time and space are produced due to interactions in a possibility space. We have now described how the universe can appear in our consciousness without a physical manifestation and disappear from our awareness without a physical destruction. The appearance and disappearance of the universe constitute the creation and destruction of *experience*, rather than a physical reality. The reality underlying this appearance and disappearance is eternal, and yet this unchanging and eternal reality seems to create a changing world. To explain these changes, we must now attribute causality to time—time is the cause of the changing strengths and powers of the information interactions.

Therefore, we can now circle back to the original question namely, that time passes because changes happen, but the underlying reality is eternal, therefore, past, present, and future exist eternally, but due to the changing interactions, they appear and disappear from our awareness. Likewise, the mechanism of time's passing—i.e. "How does time pass?"—requires a Self-Moved Mover which changes the interactions between the possibilities. What are these changing interactions? The energy and meaning required to transact messages are material. Now, *what* we want to transact this energy and meaning with requires our consciousness to be *directed* toward an object. This directedness of things toward each other—which have been called 'goals'—can be attributed to time. In simple terms, the effect of time's passing is that observers will acquire different kinds of goals; they will be able to perceive ideas that were unperceived earlier. And when these goals are formulated, then matter can be used to fulfill these goals. Or at least, the competition to attain these goals can begin with the advent of the goals.

The effect of time is thus simply to create a *purposeful* universe, in which observers can choose to participate in some purposeful activity. These purposes appear and disappear because of time, therefore, some purposes become predominant sometimes, as other purposes are subordinated. For example, deeper questions about the nature of life are becoming mostly irrelevant in modern time. These questions have been replaced by mundane queries about food, sex, and security, and instead of competing to provide a better understanding of the nature of our existence, people are competing to supply whatever they understand by better food, sex, and security.

Our world is still purposeful, but the purposes have changed, and this large-scale change in the purpose is an effect of time. If the nature of time changes, then the erstwhile dominant purposes can be revived again.

Using this basic principle, we can describe what we mean by hierarchy in causality of time—it is the hierarchy of purposes. A higher purpose defines the dynamics in a larger system, and a lower purpose controls the dynamics in a smaller system. The entire universe is therefore purposeful, and every part of the universe is thereby purposeful. That purpose is simply the presence of time—we are directed toward a future because we are directed toward a goal. So long as we exist in the universe, we cannot be devoid of purpose; if not for anything else, the competition for survival in this world will force a purpose upon us. But that force—even though it seems external—has an internal counterpart, namely, that we want to be *happy*, and survival is important because it presents us with avenues for happiness. Our goals are also driven by a deeper desire for happiness. Therefore, to say that time changes our purpose, is to say that time makes us seek various kinds of happiness in this world. Our quest of happiness is innate, but our *ideas* about happiness—i.e. what will make us happy— are byproducts of time.

As we have discussed, the intentions in matter are one of the three types of meanings. Just as time selects the conceptual and structural possibilities, similarly, it also selects the intentional possibility, and thereby, certain types of intentions or purposes become feasible at a certain time. By the feasibility of certain types of intentions, the observers get the ability to fulfill these intentions, and thereby they also come to possess them more. In effect, time creates a 'market' for happiness—there is a demand for certain types of happiness, which then pushes people to create the supply. But in this demand and supply business, not everyone gets happiness.

This is because, as we have discussed, to obtain happiness we must have the favorable effects and consequences. Hence, happiness depends on responsible choices. The material machinery of this world is a means to the goal of happiness, but it must be used responsibly. If it is used irresponsibly, then we do not get happiness. What we get is only frustrated desires.

The Meaning of "Energy Conservation"

Once we see that experiences are produced by the combinations of three kinds of meanings, 'matter' or 'energy' must be viewed semantically rather than physically. The term 'energy conservation' now has a new meaning: it means that the total number of ideas, structures, and intentions are fixed. However, since these are possibilities, therefore, their combinations are not fixed. Thus, I can use an idea in a different structure for a different intent, although not every idea can be used in every structure and for every intent. There is hence some semantic interdependence between idea, structure, and intent, which prevents random combinations from being meaningful. But barring those useless combinations, the total number of combinations of these possibilities can dramatically increase or decrease. For instance if an idea is used in contexts where it wasn't previously used, and for intents for which the structure and idea weren't previously used, then by this novel combination, new experiences will be produced, and we can say that matter is not conserved. What we are essentially saying is that *experiences are not conserved*. That there can be lesser or greater diversities in experience and thus new types of experiences can be created, and old types of experiences can be destroyed. But, underlying these experiences is an eternal possibility of ideas, structures, and intents, that remains conserved, as that possibility comprises all the possible ideas, structure, and intentions. If we call this the material reality, then it can also be said to remain conserved.

This principle of energy conservation can be distinguished from the current scientific conception of matter-energy conservation. The difference is that modern science claims that the total physical 'stuff' of the universe is conserved, whereas in semantic realism, there is no physical 'stuff'; what we consider physical 'stuff' must instead be described as experiences, and the varieties in experiences can expand or contract. The ingredients from which these experiences are produced are conserved, but the results of the combination are not conserved. More precisely, we can say that *reality* is conserved, but the *phenomena* are not conserved. However, this reality is semantic, rather than physical, so the physical principle of energy conservation—which truly pertains to the phenomena rather than the reality—must be violated. It must then be replaced by a new principle of energy conservation which pertains to a semantic reality from which the non-conserved phenomena are produced. The idea underlying energy conservation is not therefore false, in the sense that there is an objective reality that isn't created or destroyed.

In that sense, the universe is never created or destroyed. However, the phenomenal experiences in this universe are created and destroyed. These phenomena are produced due to the conscious choices of the observer, and therefore, we can say that the choices of the observers are not conserved, but the ingredients that are combined using these choices are conserved. We can distinguish these as semantic vs. physical conservation principles.

Takeaways from an Eternal Cycle

The question of whether the universe is linear or cyclical, and the answer to this question, concludes this book. The answer is that there can be overwhelming theoretical evidence for a cyclical universe, but no empirical device for cyclicality. We can summarize this chapter as follows:

- The general theory of relativity extends the ideas of special relativity into the understanding of gravitation. In the process, however, the idea of inertia, and why we *experience* a backward pressure is lost.

- To address this problem, we discussed an alternative model of causation in which the effect of gravity is not caused by matter, but by the *structure* in which this matter is organized. This approach demystifies the inability to reconcile quantum and relativity theories, because they pertain to different kinds of realities—the quantum reality is like words, while relativistic reality is like the grammar. The combination of these two things produces a world that is just like sentences; their meanings can be attributed to the words or to the sentence structure. However, since this meaning is also chosen by the observer, therefore, there is also a subjective level of meaning.

- Using this understanding of matter, we then explain the anomalies to gravity, namely, Dark Energy and Dark Matter, and show that the semantic view leads to two anomalies—namely, inertial acceleration, and non-inertial constant speed motion. These coincide respectively with Dark Matter and Dark Energy. The semantic view is therefore not merely an interpretive convenience, but it is also scientifically more useful to explain yet unexplained phenomena.

- We now begin constructing a holistic picture of reality based on the discussions of the previous chapters. This involves recognizing three kinds of spaces (or three dimensions of space), three kinds of time (or three dimensions of time), and three kinds of causation (or three dimensions of causation—results, effects, and consequences of the choices of the soul). Thus, the essential mechanism for our experience is our choice, but it operates in a changing world.

- We discussed the five-staged model of choice that involves thinking, feeling, willing, knowing, and acting, of which only the 'acting' aspect is studied in science, but because the other stages are neglected, the 'acting' stage is modeled physically rather than informationally. We describe a five-part model of informational causation through the exchanges of messages between observers.

- This discussion fully equips us to talk about the origin of the universe, and whether it is cyclical or linear: the universe can appear and disappear due to the changing strengths of interactions, and since there are two type of invisibility between visibility, therefore, the universe must cycle through the visible and the invisible states.

- This brings us to the conclusion that the cause of change is not matter, or even the observer, but time. We can all this *Causal Time* to distinguish it from parametric time, and it can be understood as that agency that creates purpose, goal, and desire in us. What is desire? It is a *method* for obtaining happiness. Time's effect is therefore that it creates diverse ideas about enjoyment or pleasure.

- Of course, this pleasure depends on responsible choices. Therefore, if enjoyment is responsible, then the enjoyment continues. But if the enjoyment is irresponsible, then it is also frustrated.

- Finally, we note a new semantic principle of energy conservation in which choices are not conserved but the ingredients that these choices combine are conserved. In a simple sense, the reality is conserved, but the phenomena created by the choices are not.

- In summary, quantum theory, general relativity, and thermodynamics pertain to three aspects of semantics—words, grammar, and intention. We can also say that these are the universal, contextual, and individual notions of meaning. If we adopt this view, then we solve all the problems of (1) unification in modern physics, (2) interpretation and understanding of these theories, and (3) explain what currently lies unexplained and hence beyond these theories. Progress in science is defined by that which brings unification, simplification, and explanation, and this approach satisfies all three.

Epilogue

> *The summer is ended and gone,*
> *How swift have the months flown away!*
> *So swift I scarce have had time to look on,*
> *And the incidents note of each passing day.*
> —Benjamin Hine

Many of the topics covered in this book have previously been part of my other books. For example, *Gödel's Mistake* discusses the issue of hierarchical space and time, and how it arises from the problems of incompleteness in number theory. *Mystic Universe* applies this idea to describe how space and time are not uniform, which leads to an alternative structure of the universe. *Quantum Meaning* discusses the problem of probabilities, entanglement, and discreteness. *Sāṅkhya and Science* elaborates on the symbolism of matter, and how it encodes information and causality is informational transactions. *Signs of Life* employs semantics to refute the idea that species evolve through random mutations, because randomness is an unsolved problem in atomic theory. *The Balanced Organization* and *The Yellow Pill* discuss the applications of hierarchical organization and the role of conflict in the evolution of organizations, institutions, market dynamics, and society's evolution. *Emotion* describes the human mind as inherently conflicted by competitive tendencies, and how these tendencies change our lives through dominant-subordinate relations. *Moral Materialism* discusses how choices have effects and consequences, and how our experience evolves as a result. *Uncommon Wisdom* discusses the role of relativity and thermodynamics, and how the pending problem of their reconciliation with atomic theory will destroy the dogmas of materialism, determinism, and reductionism in modern science. *Six Causes* discusses how the creation by God is not incompatible with science if this creation is the expansion of the whole into the smaller parts.

But I did not have a rope to tie all these ideas into a single coherent picture of reality that covers all these topics in an easy to grasp manner. *Time* is such a topic since it is present in every scientific theory. Time is personal in our experience, interpersonal in a society, and objective in every theory of science. Time covers all the issues of change, causality, continuity, reversibility, relativity, hierarchy, and cyclical evolution that appear in different ways in different subjects. By discussing the role of time, in relation to space, matter, and our experience, we can cover a much broader gamut of problems and solutions, than is possible with other types of topics. In that sense, this book is not drastically different from what I have earlier discussed in other books, but it presents the avenue for extraordinary unification.

In another sense, it also sets the yardstick for alternative scientific and philosophical doctrines, which focus on disaggregated and compartmentalized issues, never hoping or even trying to unify our understanding. What we call 'academics' today is so fragmented into various departments, that even if one department completely contradicts the other departments, the discrepancies never come to light, because the people do not work in between these departments, and seldom try to reconcile their claims. I believe that the strength of an ideology can be verified by the broadness of its applicability to natural, life, mind, and social sciences. After all, we live in one world, not in separate worlds. Our body described by natural sciences is also living, is affected by a mind, and is embedded in a society. Therefore, the body cannot be completely known unless we reconcile these subjects.

The reconciliation, however, must *begin* somewhere, and this book has chosen *physics* because there is no other subject that combines the logical and mathematical rigor of thinking with the practical and empirical verifications of these ideas. Physics forms the basis of all modern natural and life sciences. And it can speculate about the largest swaths of space and time, along with the smallest units of matter and energy. This is a fertile ground for unification because the seeds that germinate here can propagate to other sciences and fields much better than if they germinated elsewhere.

The problem is that most physicists will argue: You haven't framed the arguments with mathematical rigor, which is the *de facto* standard for physical theories. And my answer to that criticism is that logic and mathematics need an overhaul because all the foundational ideas in modern logic and mathematics are problematic, and often false. Examples of problematic ideas include the understanding of logic as the universal principles of non-contradiction, mutual-exclusion, and identity because this kind of logic must preclude all concepts and confine itself to thinking about objects. The problems of incompleteness in number theory entail that mathematics cannot be as powerful as ordinary language, and if the world necessitates ordinary language, then mathematics would also be incapable of describing the world. An example of false theories includes calculus, upon which the rest of physics rests. It is founded upon the ideas of linear, continuous, and uniform space and time, whereas the problems of modern science necessitate hierarchical, discontinuous, and non-uniform notions of space and time.

We cannot explain nature using logic, mathematics, and calculus, when the *languages* of expression are contradictory to the ideas being expressed. If we employ this language, then we will find that the language is incapable of expressing the ideas, and if these ideas were forced into the language, then we will end up with contradictions. Therefore, my discussion is anchored in ordinary language, the interpretation of facts, the generalization of scientific ideas, and bringing in greater parts of ordinary language into the ambit of science. We can broadly call this 'philosophy'. It doesn't have to sacrifice rigor and precision, but it can use words that won't fit into mathematics. If we are careful in defining these words in the same objective style that we have done in other cases, this 'philosophy' doesn't deserve ridicule.

The fundamental problem with modern science is its flawed assumptions of continuity, uniformity, linearity, reduction, and physicality. These are oversimplified dogmas, not truth. These dogmas also fail in all fields of science. But the priests of academia uphold them with their enormous influence and power. Therefore, just as science grew outside the ambit of religion, similarly, an alternative science must grow outside the ambit of conventional science. Newton had to develop calculus to frame his laws, because the existing mathematics wasn't suitable to describe change. A similar effort is also needed to frame such ideas and a new logic and mathematics is needed to present these things with precision and rigor. The problem however is that the scope of this alternative language is enormous—it must have as many powers as ordinary language. The problem now reduces to the question: What are these powers?

And why do we need them to describe nature? Efforts such as the present book try to answer that question.

We need to know what ordinary language is, why it is so powerful in describing ideas that we cannot fit into current mathematics. So, books such as this one, on the philosophy of nature, are the stepping-stones toward an alternative logic, mathematics, and natural description. If we don't read and write such books, then the subsequent steps would also not occur. So, instead of examining the ideas in this book based upon established yardsticks of precision and rigor, we must look toward new yardsticks of precision and rigor that don't suffer from the linguistic incapacities of logic and mathematics. The combination of physics and philosophy can provide the framework in which we are able to think of problems, formulate solutions to these problems, and then go after that language to express these solutions rigorously. I have no doubt these are possible, but they must come in an order, just like to reach a higher point on an edifice, we must climb a ladder. This book is hence a step forward on the ladder that leads us to the truth.

Notes and References

PREFACE

1. In Christianity, death is followed by the afterlife, but no one can come back and take birth again so, for all practical purposes, death is the end of everything on Earth for that person and time thus maintains its linear nature. In fact, the linear time in science might be connected with the predilection for it in Christianity, as science in the West was developed under the cultural influence of Christianity.

1. DOES TIME PASS?

1. In Indian philosophy, a distinction between a 'mind' and the 'intellect' is drawn. The mind is described as a 'sense' that is capable of 'intuition', just like the senses are capable of sensation. The intellect, on the other hand, is described as the instrument of rationality that operates on the memory and reasons from it. Thus, the computational idea of the mind applies to the 'intellect' and not to the 'mind'. The additional difference is that even the intellect stores the past using concepts. As a result, when we recall the past, we are often not able to put it into words. A computer on the other hand, always stores memory as words, not concepts. Therefore, there is no scenario of memory recall in computers that is not expressible into words. Indian philosophy (sometimes called Sāṅkhya) states that the intellect stops working during dreaming, but the senses and the mind remain active. Thus, the mind and the senses can perceive and conceive just like the waking experience, and yet rationality is not at work during dreaming. Hence, 'sleep' is described as one of the states of the intellect, not of the mind. The mind and the senses also become inactive during the deep sleep state; hence it is the state devoid of experience.

2. Most times when I use the term "scientific" in the text, I tend to suppose the *physicist's* view of science. This is not to ignore the possibility or existence of other kinds of sciences, but only to emphasize that the physicist's view holds a greater

sway over the other sciences than any other scientific viewpoint.

3. Carl Jung advocated the idea of a "collective unconscious" which transcends the individual conscious. But his collective reality is the expression of *commonality* of ideas, rather than an ontological claim about the objective existence of a real world of ideas outside the individual. His collective unconscious refers to the notion that we are all similar in terms of our mental constitution, which is also known as *Idealism*, and the claim is often contrasted with *Realism*, where the world is real.

4. I refrain from using the term *probability* while describing *possibility*. The difference is that a probability is understood only through a succession of events over a time duration, while a possibility can be understood in an instant as a single event.

5. The classical physical world requires that objects exert a force on every other object simultaneously. For example, the gravitational force acts at once on every object in the universe. Conscious observation, however, involves *attention*, which can be selectively directed toward certain things. This attention is neither focused on everything at once, nor can it be focused on everything at once, *in principle*. Therefore, when the observer is described materially, a fundamental discrepancy between the ability of objects to affect each other simultaneously and the ability of an observer to focus on only one thing at a time, leads to a contradiction.

6. The Greeks were troubled by the problem of knowledge as it needs two things—an external world, and a world of ideas, which leads to cognition. Plato argued that there must be a world of ideas after which the present world is modeled. The world of ideas contains the notions of a perfect chair, a perfect round, perfect beauty, and a perfect man. This perfection is incompletely reflected in this world because things in this world are mostly imperfect. Thus, the world of Platonic *ideas* was transformed into a world of *ideals*, and that simply meant giving a *definition* to everything: What do you mean by a man? What do you mean by a circle? Etc.

7. Objectivity simply means something different from the subject or observer. Objectivity is not necessarily an 'object', or something that has no 'inside'. By that definition, other people are objectively real but not necessarily objects.

8. Meditations can be practiced in many forms. Some forms of meditations involve focusing on something specific, and when the focus is drawn to a specific entity,

then it is also withdrawn from the other entities. Other forms of meditations involve the attempt to withdraw from all experiences. Generally, the meditation that focuses on something is found to be better than the meditation that withdraws from everything. Ultimately, to hold the meditation, one must be able to keep the focus for longer periods of time, and that is possible only if the object our consciousness is focused on produces happiness and destroys the unhappiness.

9. The conventional idea about a Universal Observer is that such an observer is outside space and time. The Universal Observer here is distinct from space, and outside space. But this observer is not *outside* time because the observer *is* time.

2. HOW DOES TIME PASS?

1. An assumption inherent in this ideology is that only the immediately previous events can be the causes of the next event. Events in the distant past or the future cannot therefore be considered the causes of what occurs in the present.

2. Some people might want to recognize the relativistic ideas about the subjectivity of the 'present', which is called the *relativity of simultaneity*, where different events are simultaneous by different observers. Such observers, however, are also moving relative to each other. I will return to the topic of relativity of simultaneity in a later section. Presently, we are speaking about the relativism of the *present* that would arise if there were indeed many time-objects existing simultaneously for the different observers in the same reference frame, not moving relative to each other.

3. Aristotle had a mechanical conception of the world, in which the world had to be set into motion once, by an Unmoved Mover, and thereafter the world would perpetually be in motion. Philosophy, and science, were paralyzed for two millennia by this idea, until Newton severed the idea of change in this world from the idea of what started the change. Thus, Newton's first law assumes that objects stay in a state of motion unless hindered by forces. Newton's second law then postulates that an object's state of motion is changed by a force. Newton's laws still require the existence of *initial conditions*, which is a euphemism for saying that all the particles in the world are initially put into some position and state of motion. In Newton's conception of the world, God puts the world into some state of motion initially after which the world continues to move unhindered. Whether Newton was a genuine theist, or merely accommodated God to placate the Church, is a matter

of debate. But he altered the focus of science from the questions of the Unmoved Mover (which had preoccupied Medieval Christian theologians) to the study of the *changes* to motion (taking the issue of initial motion for granted).

4. In Abrahamic religions, God creates the world *ex nihilo*, or from nothing. This is a problem because creation involves infinite amounts of energy. If God could create infinite amounts of energy, then His intervention in the world would also involve the creation of energy, and that would violate the law of conservation of energy. When energy conservation is violated, then matter is created. And when matter is created by will, then all laws of nature must become false. To avoid this problem, one could say that God only creates the world and doesn't intervene in it after the creation. But that would be contrary to religion because if God doesn't intervene in the world after the creation, then how can He provide religious or spiritual knowledge to the people in the world so that they practice religion? Thus, we are led to the tradeoff between religion and science. If nature is governed by laws, then God doesn't exist. But if God exists, then religion cannot exist.

5. Eugene Wigner authored a now famous paper entitled "The Unreasonable Effectiveness of Mathematics in the Natural Sciences", which argued that the very success of modern science in formulating mathematical laws needs an explanation. Why should nature be governed by logic and numbers? In fact, why should there by any laws of nature at all? It is entirely possible that nature did not have any laws, or least these laws need not have been the laws of numbers and logic. Our societies, for example, are governed by man-made laws. You cannot prove these laws from axioms or first principles. If God created the world, He could have arbitrarily chosen the laws. But if God did not create the world, then no laws should exist. In short, the likely outcome is that either there are random laws or no laws.

6. Alan Turing postulated the idea of a Universal Computer, which was 'universal' in the sense that it could emulate or execute an arbitrary mathematical program. In our case, the Universal Computer would be capable of computing any arbitrary mathematical law. However, the *design* of such Universal Computers turned to be problematic. In the standard design—proposed by John von Neumann—which is adopted in all computers, a Universal Computer can only execute *one* program at a time. Modern day computers try to solve this problem using "parallel processing" and "parallel instructions". Parallel processing means that while a program is waiting for some input or output, its is moved out of the CPU, and another program

is executed. Parallel instructions mean that the independent tasks in a program can be executed in parallel. For example, if you are performing three operations, such as A+B = C, D+E = F, and C+F=G, then the computation of C and F can be carried out in parallel, whereas the computation of G must be sequential. Turing computers (under the von Neumann architectures) therefore present serious problems in envisioning a Universal Computer that calculates the future states of all particles in the universe. As a result, we would view innumerable computers tied to each object a more viable hypothesis than a single computer calculating the future states of each particle. This in turn would mean that there are numerous individual notions of time, which are tied to each individual particle. Since the computation requires energy, therefore, some part of the energy of the particle must be spent in computing the next state, which means that it would be dissipated as heat. As a result, Newton's first law of motion would become false, because the particle would constantly lose energy even as it computes its next state.

7. All modern computers work based on a *clock*. The clock triggers the fetching of some instructions from memory, performing those instructions, and putting the results of the instruction computations back into memory. A faster computer requires a faster clock. A faster computer also absorbs more energy. Therefore, if the clock slows, then the energy consumption goes down. A clock with zero frequency would (theoretically) absorb zero energy and perform zero computation. Therefore, computational consumption of energy is driven by the clock. If this clock represents time, then time is the true cause of the computation of the laws. In effect, the laws are working *because* time is moving. Again, we cannot take time for granted, as a parameter that just adds to the parameters in the laws. We need some driver for the computation, and that driver is time in all real computations.

8. When the world is understood as possibilities, then the universe has far more energy than is visible at any time. For example, if there were two possible states of matter S1 and S2, with energies E1 and E2, then the classical model of 'evolution' would suggest that as S1 changes to S2, then E1 converts into E2. In short, the total energy is max (E1 | E2). But if the world is described as possibilities, then the total energy is E1 + E2, although of either of S1 or S2 is observable, then the total energy visible in the universe would only be E1 or E2. Thus, a universe of possibilities has enormous amounts of hidden energy in the sense that we cannot observe its effects as motion, but that energy exists in a potential form.

9. This is a well-known problem of Buridan's Ass who cannot decide whether to eat hay or drink water and dies due to the inability to decide. In real life, this doesn't happen because we all have a goal for survival and depending on whatever is more necessary for survival would be done first. For example, if an ass is thirstier, it will drink water, but if it were hungrier, it will chew on hay. Either way, the indecision is resolved because hunger and thirst are never equal *and* there is a goal for survival. If the goal did not exist, then one would be conflicted between alternatives because neither drinking nor eating would seem to aid in the fulfilment of the goal.

10. The relation between the self-moved mover and moved can be understood through the example of an employer and an employee. The employer is a self-moved mover because he gives the employee some objectives or goals to be accomplished. The employee is a self-mover moved because (a) he moves himself—i.e. does the work requested by the employer, and (b) his actions are driven by the employer's requests, so he doesn't bear the responsibility for the goals. Nevertheless, to the extent that the employer doesn't tell the employee how to go about doing the assigned work, therefore, the employee is a self-mover. And to the extent that the employee is moved by the instructions of the employer, he is moved. Thus, 'mover' in the case of an employer applies to the movement of the employee, and the instructions given to the employee are self-moved. Conversely, the 'mover' in the case of the employee refers to his own actions that accomplish the employer's instructions, and the 'moved' refers to the idea that he moves based on the instructions. In short, 'mover' refers to activity, and 'moved' refers to instructions. The self-moved mover moves the self-mover moved, so their causalities are closely interlinked because the first gives the instructions and the second fulfills them.

11. Causality involves predictions and explanations. A prediction requires answering the questions of 'when' and 'where'—i.e. when something will happen, and where it will happen. An explanation requires answering the question of 'why' something will happen, given that we know 'when' and 'where' it will happen. This idea is easily understood through a contrast between classical and quantum mechanics. In classical mechanics, the 'when' and 'where' are predetermined (if the initial conditions are fixed), and 'why' is the law (such as of gravitation). But in quantum theory, we cannot predict (a) when a particle will be emitted, (b) where it will be absorbed to create an event, and (c) why it is emitted at precisely that point in time and why it is absorbed in a specific place to produce an event. Hence, classical mechanics has causal predictions and explanations, but quantum

mechanics doesn't. This failure of causality requires us to rethinking causality. In the case of a causal model involving possibilities and choices, the 'space' of all possibilities assigns a location to each possibility. However, the resulting 'where' is not the prediction of an event. The prediction involves the selection of a possibility by a choice, which combines the 'where' and 'when' and hence we can call it 'what'—i.e. then event that will happen, which constitutes a goal. Similarly, to attain the goal, the observer might pass through several intermediate states at different times, which also involve a 'where' and 'when', but they are not the same as the goal. Instead, they should be understood as a trajectory to the goal. Therefore, we call this trajectory 'how'. Finally, there is an individual observer who chooses the goals and traverses the trajectory by his efforts, and constitutes the 'who', and constitutes the 'why'. In this revised model of causality, the 'where' is the unchanging space of possibilities, and 'when' is the time that selects a subset of these possibilities. Due to this selection, 'when' and 'where' must still exist in a scientific theory, although they cannot refer to the states of a given observer, but only to the states of any possible observer in the universe. Similarly, the deterministic law of motion—or 'why' something happens—still exists, but it must be supplanted by the idea of a 'who'. As a result, the number of questions in causality increase from three to six; the previous questions of 'where', 'when' and 'why' must be supplanted by 'what', 'how', and 'who'. The answers to the former questions do not settle the answers to the latter.

3. DO THE PAST AND THE FUTURE CHANGE THE PRESENT?

1. The least action principle is part of a general class of principles called *variational principles*, which assert that the correct solution to a problem involves either maximizing or minimizing some quantity. The principle originated during Greek times in Euclid's work where the 'distance' between two points was defined as the *shortest path* between them. As a result, when non-Euclidean geometry was discovered, this principle was broken in the strictest of senses. However, since the principle doesn't stipulate which quantity must be minimized or maximized, therefore, it has become possible to convert every scientific law of nature into a least action formulation, with a different definition of 'action'. The plurality of scientific theories now leads to the problem of defining what we mean by 'action' or the quantity to be minimized. In later sections, I will dispute some aspects of this idea, namely, that the *path* to a destination must be minimized, but the *goals* must be maximized.

2. Non-uniformity of space must be distinguished from the *curvature* of space, even though both represent the shortest path to a destination. The reason is that the notion of space curvature arises in General Relativity due to the presence of mass, which always accelerates a particle *if* the particle is moving *toward* the mass. A non-uniform space on the other hand is the energy barrier between a particle and a goal, that always decelerates a particle *if* the particle is moving *toward* that goal.

3. The predictions fail because kinematics only deals with the conservation of energy, momentum, and angular momentum, and each of these properties can be conserved even with numerous possible *distributions* of these properties.

4. The explanatory failure is just the converse of the predictive failure. When innumerable distributions of energy are permitted, then a state A can lead to states B, C, and D, and a state P can be caused by states Q, R, and S. The former is the failure of prediction, and the latter is the failure of explanations. These failures are also associated with the failure of *necessity and sufficiency*. For example, the state Q is not necessary to explain P, because P can also be explained by R and S. Likewise, the state A isn't sufficient to explain B, because B might not occur when A occurs; for example, when A occurs, instead of B, other states like C or D can occur. Thus, predictive success entails *sufficiency* and explanatory success entails *necessity*. If the causal explanation doesn't have necessity and sufficiency, then it is incomplete.

5. The law of gravitation, for example, defines what we mean by *gravitational potential energy*. This potential energy can be traded off with *kinetic energy*. If we know how much potential energy a gravitational field creates for a particle, then we can identify potentially infinite distributions of total energy into potential and kinetic energies. This isn't immediately evident when we study single particles, but it is evident when we take particle collections. The reason is that while considering single particles moving in a gravitational field, we don't consider the possibility that particle collisions can split and coalesce the particles. As a greater or lesser number of particles are created (i.e. the particles are not conserved; only the energy is conserved), the total energy is also distributed among these particles in different ways. The incompleteness of gravitational theory is simply that it doesn't predict whether the particles will split or coalesce and in what ways they will split or coalesce.

6. The difficulty in reaching a possible state is contextual to the state from which we are trying to reach this possibility state. In short, the mountain or valley in the possibility space must lie on the path toward the possibility. Therefore, statements about the effort required to reach a state is also subject to the state from which this effort must be made. The net effect of such contextuality is that 'equidistant' states from the destination may not require the same effort, and so all universalist claims about the effort necessary to reach a state would also become false.

7. In the previous chapter, we noted that this is because the universe is *finite*. A universe without tradeoffs would also be an infinite universe in which everything was always feasible, and one could consume forever without a shortage. In principle, the universe could afford infinite consumption, but it does not. This is because the consumption is eternally a possibility, but choices produce effects and consequences which reduce the possibilities if the choices are not responsible.

8. A variational principle relies on a calculus of variations in which some quantity is minimized or maximized. Generally, in modern science, each theory adopts only one principle—e.g. the minimization of the gravitational potential energy in the case of a hanging chain, or the minimization of the action required to cross a potential energy barrier. By 'variational principles' here I mean theories that employ more than one such principle. A typical example of such a theory is that which computes the minimum cost for the maximum value. Such theories can sometimes be stated in terms of a single principle—e.g. maximization of *profit* where the profit is defined as the difference between value and cost. In general, however, if multiple such principles are involved, we are required to explore the space of all possibilities and identify that point in space where each principle has been optimized.

4. DOES TIME PASS UNIFORMLY?

1. Such validation requires us to go a *higher dimensional* space and time in which the present space and time becomes a measured *object*. After all, to measure something, we must be *outside* that thing. Since space and time act like containers, all measurements within the container must assume the properties of that container.

2. "Spatially structured photons that travel in free space slower than the speed of light", Daniel Giovannini, et al. Science 20 Feb 2015. Vol. 347, Issue 6224, pp.

857-860. The essence of this paper is as follows: Light is generally assumed to have a *plane wave* structure—i.e. the wave oscillates only in a two-dimensional plane, and all its values perpendicular to that plane remain zero. However, light may not have such a planar structure, and this paper provides measurements on the speed of light when the light's planar structure has been changed, and the results show that as light's structure is modified, its speed reduces. The non-planar structure can be thought as encoding more *information* in the photon. Such an interpretation would indicate that the speed of light is not a constant, but that the *rate of information transfer*—i.e. how much information you can exchange in a finite time—is constant.

3. Another kind of speed of light variation naturally arises due to the Uncertainty Principle in quantum theory where if the position uncertainty was finite, then the particle would be said to be 'spread' in space. Successive measurements can in fact reveal positions far apart from each other, leading to the conclusion that light's speed is infinite for very short periods of time. This effect can be explained as not truly being a speed of light variation but the result of measurement in hierarchical space (which I will discuss shortly) where in trying to measure the 'whole' we sometimes measure the 'parts'. The 'whole' is spread, but the 'parts' are localized; however, since they are parts of the whole, therefore, we can say that we are still measuring the whole, and that inference can lead to the conclusion that light has traveled faster than its fixed speed. But I don't consider this a fundamental effect, although one that points toward a more nuanced understanding of space structure.

4. I have depicted these oscillators as one-dimensional, but that is just out of convenience and simplicity. They could as well be three-dimensional (although not more) with an innate property that indicates a directional orientation.

5. Mathematicians recognize two kinds of infinities—*countable* and *uncountable*. The countable infinity pertains to natural numbers and fractions, and the uncountable infinity pertains to real numbers. On a continuous real line, there are irrational numbers such as $\sqrt{2}$ and π, which are neither integers nor fractions, and the set of real numbers includes natural numbers, fractions, and the irrational numbers. This set of numbers is uncountably infinite, which is exponentially larger than the countable infinity. If we multiply this uncountable infinity by any integer (i.e. a countable number), the result is still an uncountable infinity. And that means that we can compress any countable number of intervals into a single

interval, and it will contain as many uncountably infinite real numbers as all those intervals together.

6. Modern crystallographic measurements suffer from this problem of simplifying to solve. For example, in a crystallographic experiment, it is assumed that light is a particle that moves in a straight line, and reflection, refraction, absorption, and passing through are occurring due to straight line particle trajectories. In a 2-slit experiment, however, each slit becomes a source of light, and all these sources then interfere to produce an interference pattern. In short, in the 2-slit experiment, we treat light as both wave and particle, but in the crystallographic experiment we treat light only as a particle. As a result, all crystallographic inferences of molecular structure are fundamentally flawed, because they ignore the quantum properties of light and treat it as a classical physical phenomenon. If quantum effects were taken into account, it would become impossible to arrive at any structure whatsoever, because innumerable arrangements of atoms can produce the same interference.

7. This model of perception can also be used to explain non-perception of the world by 'withdrawing' our senses, as is the case in meditative practices. It can be used to explain false perception or illusions, which would arise if we issue the wrong requests. Now, the illusion would be objectively real, but only because a false request is issued. For example, under fear, we might perceive a rope to be a snake, which is because we are issuing a request: "Does this look like a snake?" and the answer to that question is "Yes", because the rope has a similarity to a snake. The correct question, when we are not fearful, is: "What does it look like?", and the response to that question is: "Rope". Generally, the process of perception takes some time: we don't immediately cognize an object, which means that multiple such requests are issued one after another, and information is gathered and then interpreted. Thus, for instance, we might cognize a rope initially by asking simpler questions: "What is the shape?" and "What is the size?". Then, under fear, we might ask the question: "Does it look like a snake?" and without fear, we might ask the question: "Does it look like a rope?". The sequence of questions would also mean that if we don't possess the idea of a 'rope' *a priori*, then we will not ask: "Does it look like a rope?", and even the answer to the question "What does it look like?" would not be understood because we don't know what a 'rope' is. Hence, we must possess *a priori* knowledge of what a rope is to ask the right question and understand the answer. Thus, the request-response model explains many facts about perception.

5. IS TIME ABSOLUTE OR RELATIVE?

1. Positivism represents a radicalization of empiricism. Classical empiricists such as John Locke had argued that the mind at birth is a 'blank slate' or *tabula rasa*, and so whatever knowledge of concepts is acquired it is through everyday experience. Therefore, we don't have an *a priori* knowledge of a 'table'. We rather get the idea of a table from experience. But what about ideas such as 'color' and 'taste'? Locke believed that these are just words that we used to describe the perceptions gained by different senses. Philosophers call all knowledge gained by sense perception *a posteriori* knowledge, and positivism is the claim that all knowledge is *a posteriori*. There is, however, a key difference between positivism and empiricism, which is that in empiricism we hold that there is an externally real world, but in positivism, there is no external world, or at least we cannot talk about this reality. An example of empirical realism would be that there is indeed such a thing as objective space and time, with properties like length, duration, and direction. But according to positivism, there is no such reality—i.e. there is no objective space and time—at least we cannot talk about such a reality, because it is not amenable to observation. When positivism is applied to space and time, then the notion of an absolute space and time is dissolved, and we are only left with the perceived space and time.

2. The Twin Paradox is a thought experiment in relativity in which one of the twins sits in a high-speed rocket and travels in space for a long time. Now, each twin considers itself stationary and the other twin to be moving. Thus, according to the twin on Earth, the time has dilated for the twin in the rocket, and he should age much less than the twin on Earth. But according to the twin on the rocket, the time is dilated for the twin on Earth, and he should age much less than the twin in the rocket. The paradox is that only one of the twin must age, but by relativity each twin will think that the other one has aged, and one of them must be wrong, therefore, the principle of relativity (i.e. the equivalence of both twins) must be wrong. Now, many people argue that this is a false paradox because when the twin leaves the earth, the rocket must accelerate against gravity, and by that acceleration, the twin in the rocket will feel the increased force as the rocket leaves the Earth's gravity. Therefore, the twin in the rocket will know that he is indeed moving while the twin on the Earth is stationary. But this counterargument misses the point that the clocks and meters in the rocket have no *memory*. So, they know about the rocket's acceleration only for the time that the rocket accelerates, and not *perpetually*. Once the acceleration stops, the clocks and meters will *forget* that the rocket ever

accelerated. After that, clocks in both places behave as if there was never an acceleration. If a rocket accelerates for an hour, the twin in the rocket will age faster for that hour, not perpetually. But according to relativity both twins must continue to think that the other twin is aging—perpetually. Thus, if we give the rocket enough time, we can find that the effects of acceleration must eventually be negligible. And that would ultimately mean that only one of the twins must age more than the other. So, the paradox is real, and undermines relativity's equivalence of frames.

3. I will later argue that even these individuals are not physical; their distinctness as individuals also constitutes another kind of meaning. To understand this meaning, we can think of *persons*. A person is an individual, and even if has the same shape and size as the other persons, he still bears a unique personality, that we call their 'individuality'. I will later describe how this individuality is the different intentions or purposes in persons. To say that there are two individuals, therefore, is equivalent to saying that there are two individual purposes or intentions. This idea of individuality is semantic, and therefore, even the individuals are not physical.

4. There is a limit to the minimization of energy, as a smallest amount of energy is necessary to receive any information. Therefore, when information is growing, and energy is being reduced, therefore, we are pushing the limits of information reception in the sense that we are trying to encode greater amounts of information through lesser amounts of energy. This process requires that each packet of information must have a lower frequency and amplitude, but more such packets must be received in a shorter time. Due to the former, the energy of the transmission is decreased, and due to the latter, the total amount of information is increased.

6. IS TIME DISCRETE OR CONTINUOUS?

1. The mathematical property of orthogonality is defined by something called a 'cross product' in which two functions are multiplied and the resulting values are summed over all positions and time. If two functions are orthogonal, then the sum of products of these functions at each individual position and time will be zero. A simple example of orthogonality is two functions that do not *overlap* at any point in space or time. Due to this non-overlapping property, the value of one of the functions would always be zero and any quantity multiplied by zero will also become zero. The sum of all such quantities will then also be zero. A more complex example of orthogonality is one in which the functions do overlap but

these overlaps are *symmetrical.* For example, if one function has a constant value of 1, while the other function is symmetrical values of +1 and -1 across different locations and/or time, then the sum of these values across space and time will also be zero.

2. The principle of superposition is the assertion of the *linearity* of the system. What is linearity? It basically says that the whole is the sum of the parts. In short, if there are N outcomes, then the total outcome is the sum of the individual outcomes. Superposition is possible in quantum theory because the alternatives are orthogonal. Just as the position state of a particle is determined as the sum of X, Y, and Z positions, similarly, the superposed state is like the sum of orthogonal alternatives. The difference is simply that each of the X, Y, and Z, axes can be simultaneously measured in classical physics, but they cannot be measured so in quantum theory. Thus, superposition works only with the hypothesis of a 'collapse' in the sense that one out of the many preexisting possibilities must be seen in one observation.

3. The ideas of possibility and probability must be contrasted. Possibility says that something can happen, but it doesn't say *if* it will happen. Probability tries to assign a relative value to its happening, in comparison to other things happening. Probabilities are notoriously difficult to define, because they suppose two things: (1) that something will happen, which is not *necessary*, and (2) a notion of something happening with every unit of time, which is not *sufficient* because the unit of time can be seconds, hours, days, years, millennia, etc. The passing of time doesn't ensure that something will happen because the unit of time at which the probability is defined is not known. Possibilities are, on the other hand, very easily defined because all that is needed for possibility to exist is to explain how it can be rationally produced from something more fundamental (assuming that it is not an axiom). In short, rationality suffices to define a possibility, but empirical evidence is needed for probabilities to be correct, but no amount of time proves a probability.

4. Individuality contradicts independence whenever we consider whole-part relations. For example, my body parts—e.g. hands, legs, chest, head, etc.—are *individually* distinct things, but they are not *independent* things, because they are part of the whole body. Thus, I cannot take my legs to work, and leave my hands at home.

5. Classical physics equates the ideas of individuality with independence. A billiard ball, for instance, is an individual thing, and it is also independent of the other billiard balls, because any billiard ball can be anywhere on the billiard table, regardless of the position of the other billiard balls. This is because the collection of billiard balls doesn't constitute a whole system, or a 'body'. The equating of individuality and independence is also conceptually the statement of *reductionism*, in which there is no whole aside from the parts that comprise it. There are two ways to disprove reduction. First, we can identify the whole, but this method proves inadequate because if you call your body a whole system, then someone can say that it is merely comprised of the molecules. Second, we can say that the parts of the whole are *individuals,* but they are not *independent.* This is technically the definition of 'entanglement'. So, quantum entanglement is the counter thesis of reduction. This is often not easily recognized if we keep thinking of anti-reduction in terms of identify a boundary that separates one system from another. There is, of course, a boundary, or an ensemble, but if we reject its reality, then the problem of entanglement becomes harder, as there is individuality without independence.

6. There are many definitions of linearity and non-linearity, which are in some sense related, but in another sense different. The first definition of linearity is mathematical, namely, that an equation which has first-order variables only, is called linear. The second definition of linearity is that a system is linear if small changes in initial conditions produce small changes in the final conditions. A third definition of linearity is that an equation's variables only depend on other (first-order) variables, not on their own values. For example, if the rate of change of the variable x depends on the value of x then the equation of change is called non-linear. The fourth definition of linearity is the ability to reduce the whole to the sum of the parts, and the ability to separate the parts individually. The second and third definitions of non-linearity point toward some *inseparability* in the system—i.e. you cannot know one part of a system without knowing all the parts. The first definition of non-linearity simply extends the idea that *line is linear* so anything that doesn't follow a straight line, is sometimes called non-linear. And the fourth definition states that the system is linear because it is comprised of independent parts. Quantum theory is weird because it is linear in the fourth sense that it is indeed comprised of separable parts. But it is non-linear in the sense that no part can be defined independent of the other parts. Therefore, classical uses of the term 'linear' and 'non-linear' can sometimes become misleading because both properties exist simultaneously.

7. A classic example of multiple descriptions of a system exists in the slit experiment where light passes through some number of slits before it is absorbed by a photographic plate (or a battery of detectors). The problem is that depending on how many slits exist between the light source and the detector, the observed pattern on the detector changes. The varied patterns correspond to different ways of dividing the total energy of the wavefunction into different sets of particles. Since the same total energy can be divided into particles in many ways, therefore, the particles produced by this division cannot be considered real physical objects that exist independent of the measurement setup. These virtual particles are rather produced as a consequence of the number of slits being used during the slit experiment.

8. The principle of mutual exclusion says that if X is true, then not-X is automatically false. This principle is employed in all proofs by contradiction: we essentially prove that not-X is false to conclude that X is true. However, this principle of binary truths fails when we employ concepts. For example, if something is not a horse, it doesn't mean it is a cow. That conclusion depends on the *context*. In some contexts, where are trying to decide if something is either a cow or a horse, then mutual exclusion can be used. But in another context, the set of possibilities could be a horse and a dog, and then, not a horse would mean a dog. Thus, the principle of mutual exclusion is contextually dependent and not a universal principle.

9. The use of non-classical logic is very important due to what is known as Bell's Theorem, which says that there are not 'hidden-variable' theories of matter that can improve upon the current description of quantum phenomena. A naïve reading of Bell's theorem says that goals and meanings are hidden variables because we cannot observe them by sense perception. Therefore, the conclusion must be that quantum theory cannot be completed using hidden variables. This is a false conclusion because all hidden variables must be *physical variables*, which means that they are governed by classical logical principles of mutual exclusion, non-contradiction, and identity. Bell's Theorem rules out such physical variables, but not variables governed by non-classical logic, which are hence also non-physical variables.

10. Representation of meanings is a non-classical and non-physical idea in the sense that in classical physics, one object can never *represent* another object, as each object is independent of the other objects. However, one symbol can represent another symbol, because both symbols embody the same idea, although to varying

levels of detail and granularity. For example, a miniature replica of the Taj Mahal is a symbol of the real Taj Mahal, and it represents the original Taj Mahal. This is because both the real Taj Mahal and the replica have the same form. In a sense, the real Taj Mahal is also a symbol of the *idea* of Taj Mahal, although because this idea has a single embodiment, therefore, it is harder to think of it as a symbol.

11. The Quantum Zeno Effect was presented in a 1977 paper *The Zeno's Paradox in Quantum Theory* by B. Misra and G. Sudarshan. It describes how observing a particle preserves its state. The use of the term "Zeno's Effect" comes from the paradox presented by the Greek philosopher Zeno of Elea. In one formulation, the paradox says that to reach a distant point, you must cross half of the distance to that point. But to reach that, you must cross half of the half of the distance. Thus, if the principle is applied, it turns out that you need infinite time to move the smallest of distances, since there are infinite fractions between any two fractions.

12. A wavefunction is the combination of two things—a wave and a function. The function represents a distribution of energy in space, and the energy is represented by the wave. Essentially, in a classical sense, a wavefunction is a *standing wave* comprised of two components—one that goes from the present to the future, and the other that goes from the future to the present. The probability is the outcome of the *interference* of these two waves that removes the non-common components. In ordinary language, these two components are comprised of a forward moving goal, and a backward moving destiny. We can think of these two as money (destiny) and expense (goal). We cannot spend more than the money we have. But we can spend the money we have in different ways, suitable to our goals. The interference thus represents the fact that the goals can only be achieved within the availability destiny, although the destiny can be refactored according to the goals. Of course, the mathematics of the wavefunction is silent on these topics; it simply describes which things are *possible*, and which things have greater or lesser *probability*. These probabilities are the averaged values over many potential alternative scenarios. For example, in a certain situation, money is more likely to be spent on some things than others. Accordingly, the probability of spending changes. But the use of probabilities doesn't tell us which order in which the money will be spent, what would be the causes of that expenditure, and how that expense would be explained.

7. IS TIME REVERSIBLE OR IRREVERSIBLE?

1. We can still describe a shortest-path principle that determines the paths on the forward and reverse cycles. Such a principle would give different paths on the forward and reverse directions, but they would not be *equal* in the two directions.

2. Entropy denotes the total number of possible states of a system. Quite like atomic theory, the system can potentially be in many states. The difference is that the quantum states are the states of a system's *parts* whereas the thermodynamic states are the states of a whole system. Unlike quantum mechanics, where each state is assigned a different probability, all the thermodynamic states are equally likely. However, unlike quantum mechanics, the thermodynamic states are not orthogonal, which simply means that these states can be arbitrarily near or far to the other states. This gives rise to a probability distribution of finding the system in a state, which is called the *Boltzmann* or *Normal* distribution. In effect, even though all states are equally likely, some states become more likely due to a higher concentration of states around these states as opposed to other states. Entropy increases as the total number of states in a system increase; this is generally correlated to the increase in the number of particles, the temperature of the system, pressure, volume, etc.

3. I'm being extra cautious here in not saying that the *least action* principle is broken because least action depends on the *goal*. If we set our goal to suboptimal short-term goals, that take a longer route the destination, then the path to each short-term goal can be the best path, but the route to the longer-term goal can be suboptimal. This is important because it is possible to formulate thermodynamics also in terms of least-action principles, and what that really means is that we can take the most optimal route to the short-term suboptimal goals. In the truest sense, the least-action principle is violated if we consider the possibility that we might want to optimize the paths to the end-goals, rather than the paths to intermediate goals. But the principle may not be violated for the short-term goals and paths.

8. IS THE UNIVERSE ETERNAL OR CYCLICAL?

1. This indeterminism in relativity came to be known as the 'Hole Argument'. The essence of this argument is that the total mass and energy of the universe is compatible with innumerable alternative matter distributions. Of course, this was also

true in Newton's physics, and each particle had to be set into a specific state at the beginning of the universe. However, as we have seen previously, setting the state of the universe once doesn't fix the state subsequently because collisions of particles can redistribute the masses and energies of the particles. In short, the energy and mass are *conserved*, but the total number of particles is not conserved. This means that for the same total mass and energy there can be a greater or lesser number of particles. The novelty is simply that the indeterminism in relativity *grows* because now energy can be attributed both to gravity and electromagnetism.

2. Even in the gravitational scenario, the energy is not 'gained' but converted from potential to kinetic. The fact, however, is that we don't know that there is gravitational potential energy. We just know that the accelerating object gains *kinetic energy*, and the law of conservation of energy is then used to say that the kinetic energy must have existed previously as a potential energy. However, what form does this potential energy take? We could also say that this potential energy existed in the object as ability to move off its own accord, so the principle of energy conservation would not be violated. However, the nature of the potential energy would be transformed from gravitational potential to another kind of potential, quite like when we pull or push things, a previously invisible form of energy becomes visible.

3. You might say: by Occam's Razor, two is better than three, and under the present circumstances, where we take gravity to be a complete theory for determining cosmological models, that is indeed true. However, I will shortly describe why gravity fails and to account for this failure, two additional physical constructs—called 'dark energy' and 'dark matter'—are postulated. Now, we can say that one is better than three, and, if there was a single mechanism that could account for gravity, dark matter, and dark energy, then even by Occam's Razor, that single mechanism would be preferred. The so-called 'rope' in this case is aimed at substituting not the objects, but the gravitational force, and at present, it is a 1-1 substitution, and the two could be considered 'equivalent'. However, when we add dark matter and dark energy to this problem, then the single explanation is much preferred.

4. This statement should be qualified in the context of modern science that employs mathematical laws. A mathematical law of the form A = B also has a grammatical structure, in which A and B are like noun-phrases, and the sign '=' is the equivalent of the verb 'is'. We also use words like 'energy' and 'position' which are concepts, and we use numbers which *refer* to the real world. Thus, you can argue that

concepts, structures, and references exist even in mathematical equations. In what way is that any different from the structure in the sentences? The short answer to that question is that a mathematical equation is not a physical object. In fact, according to science, it doesn't exist materially; it either exists in our 'minds', which are outside the material world (due to the mind-body duality), or they exist in a Platonic world of ideas which are by definition beyond this world. For example, gravity doesn't act on the gravitational mathematical equation, and the equation doesn't exert a gravitational force on the world. When I talk about the world being sentence-like, I am referring to the presence of sentences and equations in this world, and they being capable of interacting with the world in the same way as we suppose other material objects do. This simple problem then compels us to think of even the material world in terms of concepts, that can refer to other objects, and the structure of equation or the sentence is indeed the structure of those objects.

5. This view of the difference between quantum mechanics and relativity theory immediately demystifies the question of why science is unable to reconcile these two theories. The answer is that they are factually not the same. Quantum theory is the theory of how complex words are formed from alphabets (or phonemes, because each phoneme has an innate meaning—this idea is called *phonosemantics*), and relativity theory is the theory that describe how these words combine to form sentences. Their distinction pertains to two distinct kinds of meanings attached to symbols—a conceptual and structural meaning, which we generally call the distinction between word-meanings, and figures of speech, attached to the words.

6. For this reason, I had to postpone the discussion of general relativity after the discussion of quantum theory, whereas the discussion of special relativity must precede the discussion of quantum theory. The simple issue is that we need to understand that physical interactions are occurring using 'messages' rather than 'forces', to say that light is a message carrying meaning. Once we establish that physical interactions are messages, then we can also say that the objects exchanging these messages must also be meaningful objects. As a counterexample, computers at present exchange messages, and are programmed using human readable languages, but computers are physical entities. Therefore, the idea that an object exchanges messages is a necessary condition to think of the objects under study as semantic entities, and the former builds upon the latter. However, once we understand that the objects themselves are semantic, then we can talk about the structure within the objects and the messages, because the distinction between 'matter' and 'force'

is now collapsed in the regular sense of words used in modern science—these are simply encodings of meaning, although some encoding of meaning is like an object and an observer, and some encoding is like the messages between them. It is possible that the problem and the solution could be described in another way, but I found this method progressive, although it creates the difficulty that special and general theories of relativity must be intermediated by quantum theory.

7. The Dark Matter hypothesis supposes that there is additional mass, which is invisible, which then exerts a larger tug at the outer regions of the galaxy, which then cause these regions to rotate faster than they should according to gravity.

8. In principle, it is possible to say that the leptons and quarks in another part of the universe have different masses and charges, and this changes the sizes of the atoms, and the light spectrum they emit. Since these mass and charge constants are free floating values—i.e. there is no theoretical contradiction in saying that nature permits different values of mass and charge—therefore, these values can be permitted. However, the implications of this claim are serious, because now we would have to explain the values of each such constant based on the location in space. This will wreak havoc on all classical ideas of space and time where the universe is uniform and isotropic, and upon this uniformity and isotropicity rest the laws of conservation of energy, momentum, and angular momentum. Thus, if we say that the universe is not uniform, then the entire edifice of physical science will come crashing down, as simply in exchanging energy between different galaxies, the energy would not be conserved, and that would make all physical theories false.

9. A classic example of such feasibility is the evolution of species, the emergence of societies and cultures, or the prevalence of different kinds of objects and ideas at different times. The past, present, and future are different in the sense that some possibilities that were visible in the past are not visible in the present, and what is visible in the present may not be visible in the future. As pure possibilities, every-thing (e.g. all species, cultures, societies, objects, and ideas) are eternal. But they occasionally become prevalent, and sometimes disappear from observation.

10. State machines are commonly used in all computer interactions; this is a vari-ation of the standard state machines employed in computers which are based on a pair of 4-tuples: {current state, event, next state, action}. The 'action' in these state machines includes sending a message, and the 'event' is receiving a message. These

messages contain some information, which can include the sender's state, and some requested information or action. A program that executes this state machine is separate from the state machine in computer programs, and it is part of the state machine in our case. This is because when we consider the real world, then the conventional distinction between 'data' and 'program' is invalidated; the data is also received as an action, and program also produces some data. Barring some of these key differences, the interaction model is quite like that used in computers.

Index

www.ingramcontent.com/pod-product-compliance
Lightning Source LLC
Chambersburg PA
CBHW021350150726
47989CB00005B/2190